Vedic
Mathematics

*A Mathematical Tale from the
Ancient Veda to Modern Times*

Vedic
Mathematics

A Mathematical Tale from the
Ancient Veda to Modern Times

Giuseppe Dattoli
Silvia Licciardi
Marcello Artioli
ENEA, Italy

World Scientific

NEW JERSEY • LONDON • SINGAPORE • BEIJING • SHANGHAI • HONG KONG • TAIPEI • CHENNAI • TOKYO

Published by

World Scientific Publishing Co. Pte. Ltd.

5 Toh Tuck Link, Singapore 596224

USA office: 27 Warren Street, Suite 401-402, Hackensack, NJ 07601

UK office: 57 Shelton Street, Covent Garden, London WC2H 9HE

Library of Congress Cataloging-in-Publication Data

Names: Licciardi, Silvia, author.

Title: Vedic mathematics : a mathematical tale from the ancient Veda to modern times /
Silvia Licciardi, Giuseppe Dattoli, Marcello Artioli, ENEA, Italy.

Description: New Jersey : World Scientific Publishing, [2021] |
Includes bibliographical references and index.

Identifiers: LCCN 2021006490 | ISBN 9789811221552 (hardcover) |
ISBN 9789811221569 (ebook for institutions) | ISBN 9789811221576 (ebook for individuals)

Subjects: LCSH: Hindu mathematics. | Mathematics, Ancient--India.

Classification: LCC QA27.I4 L53 2021 | DDC 510.954--dc23

LC record available at https://lccn.loc.gov/2021006490

British Library Cataloguing-in-Publication Data

A catalogue record for this book is available from the British Library.

For any available supplementary material, please visit
https://www.worldscientific.com/worldscibooks/10.1142/11858#t=suppl

Dedicated to Franco Ciocci (1955–2018)

Contents

Preface

I have always had (and continue to have) a prejudice towards everything that is dressed with clothes flavored of mysticism and referring to ancient doctrines. When I was young there was a flowering of movements inspired by ancestral wisdom (true or presumed) and many embraced these movements, looking for ancient purity. Many (the most) came out, more or less unscathed, but of many I have no more news. I was immune, not by thoughtful belief but by epidermal prejudice, and this remained for the almost seventy years of my existence and, I hope, for what remains of it.

The only time I had a close encounter with one of the bearers of ancestral wisdom, was towards the middle of the 70s of the 20th century. It was summer and hot in Rome, if I remember correctly it was June (or perhaps July) of '75. I had looked hard for my doctoral thesis subject, with the hope of getting an interesting topic useful for a post-doc position somewhere in Italy or abroad. At the end I got it and I was going back home, from the University. I was carrying a heavy folder, full of articles, which I had photocopied at the Physics department (no stick pen or internet at the time!), regarding the thesis subject. I was either excited and worried by the amount of new stuff I had to learn. I was upset because I was going on foot. I had paid the copies of the articles and I had been left with no money for the bus ticket. My mood was not the best. I had arrived, not far from home, in Piazza San Giovanni, flooded by a splendid light despite the unbearable heat. I had thrown myself on a bench to catch my breath and put my nose on the cards to see which article was easier to understand. A boy of my age, no more than 22 or a little older, dressed in a white tunic with a shoulder bag, like that one of the bonzes, approached me and asked me in Italian, with a strong American accent, what I was doing. The heat, the worries for what I should have done, my natural prejudices and everything else did not put in the best of mind and I invited him to look after his business. The boy didn't get upset and told me that I was so rude just because I was a materialist and the study of neutrons and protons wouldn't have me made better. Evidently he had peeked over my papers. He wished me that the truth, that of the ancient Vedic texts, could illuminate me and after having sprinkled me with fragrant incense, he walked away calmly. That's all!!! It is just an episode, which left no significant traces, if not the adjective: *"Vedic"*.

The years passed and I got the doctorate, I had a job that had to do with the numbers and the abstractions of mathematics. I learned to dissect formulas and concepts and glimpsed a sort of spirit, almost magical, that inspired or perhaps guided intuitions of the Great Mathematicians. Among those, the greatest, belong to a privileged category of MATHEMAGICIANS, sometimes in conflict with the mathematicians themselves. Euler, Ramanujan, Feynman, Bernoulli, Heaviside, Riemann, Glaisher, Burchnall... they have opened new fields of research of incomparable wealth and elegance. Their great abilities lay in creating math from nothing, they are the quintessence of math itself. They are not to be confused with neo-Platonic, recurring enemies of science. I have witnessed, with my great disappointment, at the birth of neo-Platonic research trends of "guide numbers", to a newer-alchemic need which, although not declared, seems to occupy the dreams of many scientists. I met two traveling companions: an Engineer, *Marcello Artioli* and a Mathematician, *Silvia Licciardi*, that will help me say this and more.

Some years ago I was in India, I had been invited for a conference. My friend, Prof. *Vivek Asgekar*, at the time dean of the University of Pune, gave me a present, a book entitled:

"Vedic Mathematics"

"Vedic" - "Magic"!!... and the circle was closed. The old wish seemed to come true through that reading. That was not the case. Yet something has emerged, the reading has been interesting, I have learned that "Veda" means "Truth" and I had so much fun and, "damn math", me, my math friend and my engineer friend will try to explain why.

Giuseppe Dattoli

Frascati, September 2020

Warnings and Disclaimers

Before entering the specific topics of this book, we think it is useful to clarify the (narrow) cultural background of the Authors.

Giuseppe Dattoli is a physicist who has worked, for more than forty years, on the physics of lasers and accelerators. He is also quite geared towards Mathematics and is an expert on Operator Theory and Special Functions.

He is not an expert of anything else.

Silvia Licciardi is a mathematician with a strong background in Symbolic Calculus, with some competence in Physics, fascinated by the Philosophy of Science but...

... She is not very much versed in it.

Marcello Artioli is an engineer who has always worked in Computer Science, he is interested in Epistemology and, in a narrower sense, in the Philosophy of Science.

Unfortunately he is not even an expert on those.

It can be said that the authors are sufficiently informed on the facts, concerning Physics, in its general aspects and Mathematics, also for what concerns its historical development. Their competence in Philosophy is just enough to support a discussion of a reasonable cultural level, but it is particularly inadequate if we talk about Eastern doctrines. They are therefore not experts in Vedic doctrine, not even superficially.

This clarification is necessary because the only encounter with this topic originated from the episode outlined in the preface and from the reading of a book entitled "Vedic Mathematics" (Motilal-Banarsidass Publishers, Private Limited, Delhi, 2003). Frankly speaking it was not easy even to understand who the author is but apparently the text is based on the work of *His Holiness-Jagaduru-Sankaracarya Sri Bharati-Krsna-Thithaji-Mahaaraja* entitled

"Sixteen Simple Mathematical Formulae derived from the Vedic Doctrine".

The transcription of the author's name is a transliteration from Sanskrit and is inevitably approximate and has not been reported correctly, for obvious incompetence, with the right accents and other phonetic notations.

The transliterations are a perverse topic and an illustrious mathematician, Philip J. Davis, in his book[1] *"The Thread: A Mathematical Yarn"* has narrated a sort of surreal journey between culture, irony and mathematics that started from transliteration from the Cyrillic of the name of the Russian mathematician (Чебышёв) transliterated in western characters as Chebychev, Chebysheff, Chebyshov, Tschebyshev, Tchebycheff ... Tschebyscheff. Around any transliterated form, a sort of school of thought had flourished and the various exponents had transformed it into a religious truth. All this unbeknownst to Davis, who was one of the first to use the interpolation techniques (of Чебышёв) for the Calcomb graphics programs. In presentation of his work he used one of the forms (we do not remember which) of transliteration and

Open up Heaven !!!

---

[1]Springer Verlag eds, ISBN-13:9783764330972, ISBN: 376433097X, 1983.

This book takes its cue from Vedic Mathematics to walk around certain aspects of Mathematics usually not treated during university courses. It is a tour between the facetious and the serious, not always elementary. If we wanted to make a comparison with the mountaineering terminology we could refer to a fourth degree, whose level of difficulty is synthetically expressed as it follows:

"There are few footholds and supports, a good knowledge of the climbing techniques is required along with a specific physical training".

Apart from metaphor, knowledge of the calculus is required (derivatives, integrals, series ...) and a good disposition in "the art of computing". We will try whenever possible to make the discussion self-consistent but, for the moment, we are not entitled to perform miracles.

Any useful suggestions and comments are welcome. The Authors can be contacted at:

pinodattoli@libero.it
silviakant@gmail.com
marcello.artioli@enea.it

Chapter 1

Mixing Up Ancient and Modern

1.1 Introduction: Pythagoras' and Euclid's Theorems

The Indian mathematical doctrines, as well as the Assyrian-Babylonian, have an undoubted cultural charm, because they do not seem a self-contained body, built on Axioms and Theorems (as in the Greek tradition), but arise as one set of insights to be introduced ad hoc in the context of specific needs. The Pythagorean Theorem, as presented in the Elements of Euclid, requires a long elaboration and is introduced after the study of the properties of triangles, of Euclid's Theorems ... The Indians conceived it in a hybrid form (horrifying for the Greeks) but effective from a practical point of view[1].

With reference to Fig. 1.1, we note that the surface of the inner square (Q_{in}) is linked to that of the outer square (Q_{out}) by the obvious identity

$$Q_{in} = Q_{out} - 4\,T \qquad (1.1.1)$$

where T represents the area of the triangles with edges a, b. The

---

[1]There is an astonishing number of "independent" proofs of the Theorem. A. Bogomolny reported 122 different demonstrations in http://www.cut-the knot.org/pythagoras/index.shtml.

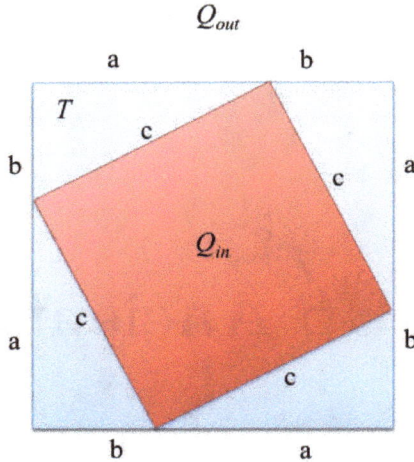

Figure 1.1: An algebraic proof of the Pythagorean Theorem.

surface of the outer square is

$$Q_{out} = (a+b)^2 \tag{1.1.2}$$

and being

$$Q_{in} = c^2, \qquad\qquad T = \frac{ab}{2}, \tag{1.1.3}$$

it follows

$$c^2 = (a+b)^2 - 4\left(\frac{ab}{2}\right) \tag{1.1.4}$$

thus eventually yielding the identity

$$c^2 = a^2 + b^2 \tag{1.1.5}$$

which once a and b are known, specifies the edge c. It represents the **Pythagorean Theorem**.

An Euclidean Bourbakist[2] would have good reasons to object, because the rules of the game were not respected. We have indeed

---

[2] *Bourbakist* is an adjective indicating a mathematician belonging to a school of extreme mathematical rigor. The name Nicolas Bourbaki is the collective pseudonym chosen by an authoritative group of French mathematicians, who in-

forgotten to prove that the four triangles of sides a, b are equal and that are right angled, we have not respected the rule according to which the use of algebra in a geometric demonstration is contrary to the spirit of the geometrical proof itself... Despite these shortcuts, the method is undoubtedly effective. Perhaps it was an intuition of this type that brought Pythagoras (Assyrians and Indians as well, before him) to formulate the Theorem in 6th century BC.

We could take advantage of the same procedure to deduce the well known ***Euclid's Theorems***.

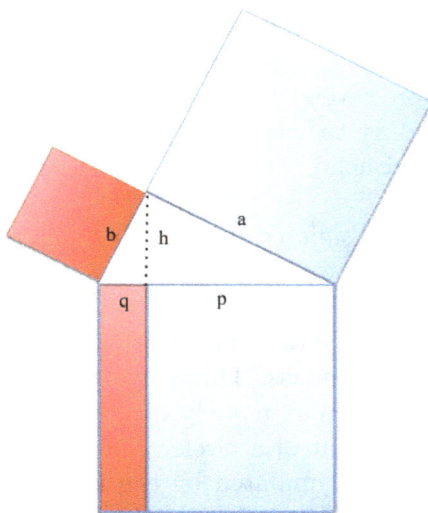

Figure 1.2: Euclid Theorems.

In Fig. 1.2 we keep the height relative to the c side of one of the triangles inside the square of the Fig. 1.1 and, denoting by p and $q = c - p$, the projections of the catheti a and b (see the figure) on the hypotenuse c we obtain

$$h^2 + p^2 = a^2, \qquad\qquad h^2 + (c - p)^2 = b^2, \qquad (1.1.6)$$

tended to reformulate the teaching of Mathematics on new grounds which accepts only a well organized system of Axioms, namely the opposite of what we have discussed so far.

which hold as a consequence of the fact that the triangles of edges (h, a, p), (h, b, q) are right angled. Subtracting the respective sides of the two previous equalities, we arrive at the following result

$$2cp - c^2 = a^2 - b^2 \qquad (1.1.7)$$

which once embedded with the Pythagorean Theorem yields

$$cp = a^2, \qquad\qquad cq = b^2. \qquad (1.1.8)$$

Furthermore, once combining the previous identities, we end up with

$$h^2 = a^2 - p^2 = (c - p)p, \qquad\qquad h^2 = pq. \qquad (1.1.9)$$

The last two identities are the Euclid's first and second Theorems, whose meaning is illustrated in Fig. 1.2.

We have therefore reversed the point of view expressed by Greek mathematics according to which the Pythagorean Theorem is deduced from those of Euclid.

The ancient Greeks suffered from many idiosyncrasies. With regard to geometry, they fixed precise rules, which determined the genesis of the great questions that lasted for millennia. According to Greek mathematicians, the tools allowed for geometric constructions were the ruler and the compass. This prescription prevented the possibility of squaring the circle, namely of constructing a square with the area equivalent to that of a circle. The proof that π is a transcendental number was stimulated by the impossibility of squaring the circle according to the ancient Greeks prescription. The solution of the problem required more than two thousand years. Entire new fields of Math had to be explored, astronomical distances in math knowledge had to be covered, before stating that circle cannot be squared employing Euclidean tools only. The answer was contained in the transcendental nature of π, namely that it is not the solution of any algebraic equation with rational coefficients. A statement so apparently conceptually far from the formulation of the problem it solved. This is a funny aspect of Math, it happened also for the solution of the last Fermat Theorem, whose solution required the creation of new chapters in mathematical thought, as we will comment in the forthcoming parts of this book.

1.2   Numbers and Geometry

The ancient Egyptians too knew Pythagoras' Theorem before Pythagoras himself. They transformed it into a technological tool.

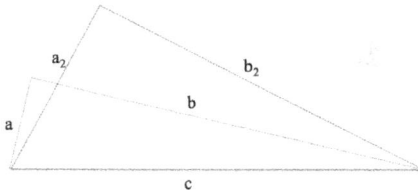

Figure 1.3: Egyptian or Pythagorean triad for $a = 16, b = 63, c = 65$ and $a_2 = 33, b_2 = 56, c_2 = c = 65$.

According to Fig. 1.3 a cord can be divided by a number of equidistant knots, say 12, so as to form three consecutive groups of $3, 4$ and 5. This triad constitutes the so-called Egyptian or Pythagorean triad and, since

$$3^2 + 4^2 = 5^2, \tag{1.2.1}$$

the nodes mark the sides of a right-angled triangle and therefore can be used to "square" blocks of stone to be used in constructions.

The Assyrian-Babylonians had gone further: a concrete proof of their knowledge is codified in their texts on clay tablets (see the Fig. 1.4 below).

Documents, dating between 2000 and 1500 BC, testify that they knew the properties of the cords in a circle, the volume of the pyramid and ... the Pythagorean Theorem. Regarding the latter, they listed "tables", on clay tablets, which described Pythagorean triads, or three integers that satisfy the Pythagorean Theorem. The titanic computational effort (for the times) had been largely justified by the practical outcome. Today we would not need it, but we could also ask ourselves what links the Egyptian triad to others, such as those

Figure 1.4: Clay tablet.

listed below

$(3, 4, 5)$	$(5, 12, 13)$	$(7, 8, 15)$	
$(7, 24, 25)$	$(12, 35, 37)$	$(15, 36, 39).$	(1.2.2)

To understand this link, we write our triad $(3, 4, 5)$ as a column vector

$$\varepsilon = \begin{pmatrix} 3 \\ 4 \\ 5 \end{pmatrix} \qquad (1.2.3)$$

and ask ourselves the problem of seeking a suitable transformation which by acting on the vector "returns" another Pythagorean term, for example

$$L \cdot \varepsilon = \begin{pmatrix} 5 \\ 12 \\ 13 \end{pmatrix}. \qquad (1.2.4)$$

By skipping the various, not entirely trivial, aspects of the search for the solution, it is possible to prove that this transformation exists and that can be expressed in terms of a matrix, given by

$$L = \begin{pmatrix} 1 & -2 & 2 \\ 2 & -1 & 2 \\ 2 & -2 & 3 \end{pmatrix}. \qquad (1.2.5)$$

In more general terms it can be proved that the solution is not unique and other two (independent) matrices accomplish the same task, namely

$$U = \begin{pmatrix} 1 & 2 & 2 \\ 2 & 1 & 2 \\ 2 & 2 & 3 \end{pmatrix}, \qquad\qquad R = \begin{pmatrix} -1 & 2 & 2 \\ -2 & 1 & 2 \\ -2 & 2 & 3 \end{pmatrix} \qquad (1.2.6)$$

The previous matrices are called **Hall matrices**[3], they were introduced in the early 1970s of the last century as the result of a non-trivial effort, which demonstrates how a problem, several thousand years old, has continued to arouse interest (and not only historical) until recent times.

1.3   Algebraic Equations and Geometry

The nuisance of the Greeks for everything that could not be constructed with ruler and compass has already been underscored. While they recognized that an irrational number, like $\sqrt{2}$, can be constructed with Euclidean tools (as shown in the following figure), they badly accepted its irrationality: a stain in the Pythagorean universe. It was kept as a secret, in the circle of initiates, and those spreading it out were punished with death!

The construction of this irrational number involves a fairly simple procedure (see Fig. 1.5):

1. Define an oriented axis, that we will say the line of the numbers, fix on this a point A that stands out 1 from the origin O;

2. Construct a segment BA of length 1, perpendicular to the line;

3. Determine the length of the OB segment, which we know to be $\sqrt{2}$ as a consequence of the Pythagorean Theorem;

---

[3]See e.g. A. Hall, Genealogy of Pythagorean Triads, Classroom Notes 232, The Mathematical Gazette, vol. 54, 390, 1970, pp. 377–379 (and reprinted in Biscuits of Number Theory, editors Arthur T. Benjamin and Ezra Brown) or B. Berggren, Pytagoreiska triangular, Tidskrift for Elementar Matematik, Fysik och Kemi, 17, 1934, pp. 129–139. For a more recent study see J. Miki, A Note on the Generation of Pythagorean Triples, MAT-KOL (Banja Luka), XXIV, 1, 2018, pp. 41–51 www.imvibl.org/dmbl/dmbl.htm, doi: 10.7251/MK1801041M.

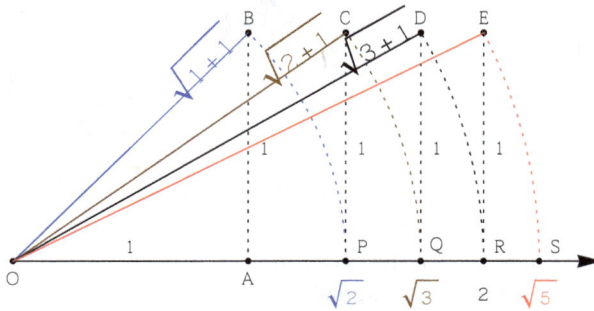

Figure 1.5: Geometrical constructions of some irrational numbers.

4. The point B can be projected on the axis OA by carrying a rotation pointing the compass at O, we have determined the position of $\sqrt{2}$ on the line of numbers;

5. This procedure has allowed the construction of a "quadratic" irrational number using the means of the Euclidean paradigms.

We can push the argument even further, a quadratic irrational can be viewed as the root of a second degree algebraic equation. The just foreseen simple procedure can be extended to display the geometrical nature an equation of second degree.

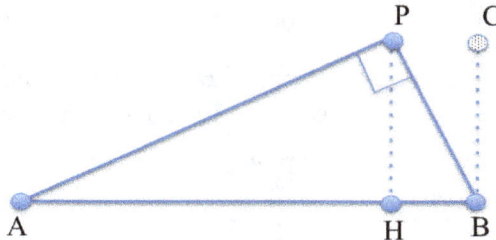

Figure 1.6: Geometrical point of view of a second degree algebraic equation.

To accomplish this task we refer to Fig. 1.5 and follow the instructions given below.

1. Build a segment, bounded by the ends A, B;

2. Construct a segment of length $< \frac{AB}{2}$, perpendicular to AB at one of the extremes, say B, and indicate the upper end with C;

3. Construct a semicircle of diameter equal to AB;

4. Move BC, parallel to itself, inside the circle, indicating with P the point of contact with the circumference and with H the point lying on the diameter AB, so that $BC = PH$;

5. From elementary geometry we know that the APB triangle is right angled;

6. Apply the second Euclid Theorem and get

$$AH \cdot HB = PH^2 \tag{1.3.1}$$

7. Use the following definitions

$$AH = x \qquad HB = AB - x \qquad AB = s \qquad PH = p \tag{1.3.2}$$

8. Then get

$$p = \sqrt{xs - x^2} \tag{1.3.3}$$

   which is a second degree algebraic equation, with a transparent geometrical meaning in the spirit of the ancient Greek mathematics.

Quadratic irrationals are accordingly nicely fitted within the geometrical cage.

Cubic irrationals are even more scandalous than their quadratic counterparts. The number $\sqrt[3]{2}$ cannot be constructed with ruler and compass. That $\sqrt[3]{2}$ is irrational it is easy to prove it, in light of what the Greeks did not know. If it were rational, we would be authorized to write

$$\sqrt[3]{2} = \frac{m}{n}, \quad \forall m, n \in \mathbb{Z} \qquad (1.3.4)$$

according to which, after taking the cube of both sides, we find

$$m^3 = 2n^3 = n^3 + n^3 \qquad (1.3.5)$$

and we might accordingly have stated the existence of a triple (m, n, p) satisfying the identity

$$m^3 = n^3 + p^3 \qquad (1.3.6)$$

In clear contrast with the Fermat Theorem (not important if $p = n$). Therefore, through the "reductio ad absurdum" procedure we have shown that $\sqrt[3]{2}$ is an irrational number (a nice exercise for the reader is to check whether the same method could be used to prove that $\sqrt[n]{2}$ is irrational ...).

Let's ask ourselves if there is a "super-Pythagorean" triangle whose sides satisfy for example the identity (see also[4] Fig. 1.7)

$$a^3 + b^3 = c^3. \qquad (1.3.7)$$

The use of the algebraic rule of the decomposition of the sum of two cubes

$$a^3 + b^3 = (a + b)\left(a^2 - ab + b^2\right), \qquad (1.3.8)$$

eventually yields

$$c^3 = (a + b)\left(a^2 - ab + b^2\right). \qquad (1.3.9)$$

Albeit "Super-Pythagorean" triangle, it satisfies the Theorem of cosine[5]

$$c^2 = a^2 - 2\,ab\cos(\phi) + b^2 \qquad (1.3.10)$$

---

[4]It should be noted that numbers representing the sides of a triangle are positive and satisfy the triangular inequalities, namely $a + b > c$. Under these assumptions the Fermat Theorem is straightforwardly proved using elementary means.

[5]Sometimes defined as Carnot Theorem, it was however well known to Euclid.

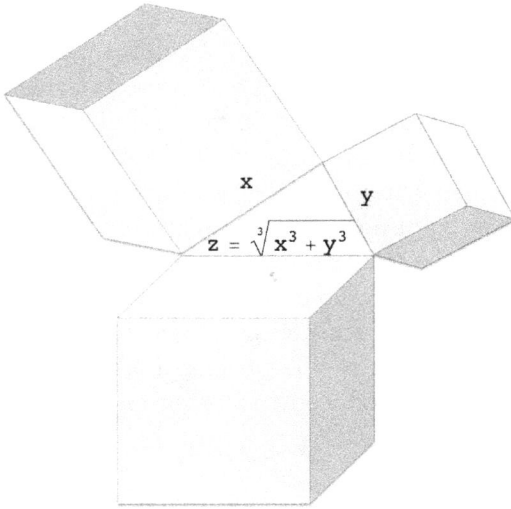

Figure 1.7: Super-Pythagorean Theorem.

which, combined with the fundamental identity for Fermat triangles, yields

$$\frac{c}{a+b} = \frac{a^2 - ab + b^2}{a^2 - 2\,ab\cos(\phi) + b^2}. \tag{1.3.11}$$

By assuming that the triangle is isosceles with $a = b = l$, we reduce the above identity to

$$\frac{c}{2l} = \frac{l^2 - l^2 + l^2}{l^2 - 2l^2\cos(\phi) + l^2} = \frac{1}{2 - 2\cos(\phi)} \quad \Rightarrow \quad \frac{c}{l} = \frac{1}{1 - \cos(\phi)}. \tag{1.3.12}$$

On the other side, it is also true that

$$l^3 + l^3 = c^3 \quad \Rightarrow \quad \left(\frac{c}{l}\right)^3 = 2 \tag{1.3.13}$$

which yields

$$\sqrt[3]{2} = \frac{1}{1 - \cos(\phi)}. \tag{1.3.14}$$

Therefore, if we construct an isosceles triangle of side 1 (with an angle at the vertex of about $\phi = 1.364\ rad$) we have automatically obtained a super-Pythagorean triangle and we have geometrically constructed the irrational number $\sqrt[3]{2}$ ... Problem solved? Obviously not!!! At least according to the prescriptions of the ancient Greeks. In the construction we did not use either row or compass and we made everything depending on the solution of an equation of third degree, unknown to the Greeks and not-soluble with Euclidean instruments. The ignorance of the algebraic equations higher than the second caused the Greeks not a little trouble. During a plague that raged in Athens, the Delo oracle ruled that the scourge would be removed if the volume of the temple dedicated to Apollo had doubled. It was a cube with a side of length l. In modern terms the problem of the oracle is reduced to the solution of the equation

$$x^3 = 2\,l^3 \tag{1.3.15}$$

where x is the side of the cube, doubling the original volume. The problem, *fatally* (it must be said), is that of defining the cube root of 2, which, as we have seen, was not easy for the Greeks. The solution frustrated the best minds at the dawn of Western thought and the plague further raged. A mathematician by the name of Menecmo (about 320 AC) proposed a solution of geometric nature, through the use of conics. Let's try to follow an argument that makes us understand how the problem can be solved in these terms.

We rewrite our equation in the form

$$\frac{x^2}{l} = \frac{2l^2}{x} \tag{1.3.16}$$

then we set

$$y = \frac{x^2}{l} \quad \Rightarrow \quad y = \frac{2l^2}{x}. \tag{1.3.17}$$

The intersection between the parabolas and the hyperbola will provide the value of the unknown x. Even combining the previous relationships we can get another parabola

$$y^2 = 2\,lx \tag{1.3.18}$$

which intersected with one of the previous conics will provide the same solution, as it has been shown in Fig. 1.8.

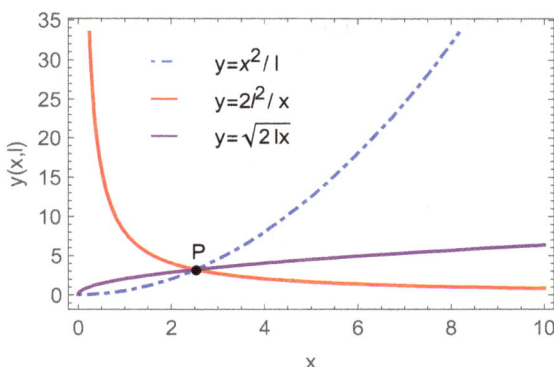

Figure 1.8: Intersection between the parabolas (1.3.17) and the hyperbola (1.3.18) for $l = 2$. $P \equiv (2.5198, 3.1748)$ is the intersection point.

The problem was solved, other proposed solutions (even one by Plato) will not be reported here. The chronicles do not tell if the plague had diminished or not. We could, however, venture a (modern) hypothesis: infectious diseases can be described in terms of the predator-prey mechanism, the predator in this case is the bacillus of the plague and prey is man. From a mathematical point of view, after a long enough time, the system reaches a state of balance and the disease recedes. If the time necessary to find the solution was comparable to that characterizing the dynamics of the predator-prey interaction the Greeks might have convinced themselves that it was just the strength of the conics, to defeat the raging morbus. In addition we have not too much arguments to say that they were not right. It may be argued that Menecno did not use Euclidean tools, but conics. The rebuttal is straightforward, conics too can be constructed with ruler and compass. In the 18th century, the Italian mathematician, Lorenzo Mascheroni showed in his book "La Geometria del Compasso" that the compass was sufficient to accomplish the Euclidean geometric construction. One of the problem treated in the book was even that of duplicating the cube by the use of compass.

1.4   Babylonians and Second Degree Algebraic Equations

Since we have quoted the equations of second degree and the attitude of the people of pre-Hellenic culture to have a more flexible attitude towards Mathematics itself, to which they looked for practical reasons, we believe it is appropriate to say something about the algebraic abilities of the Babylonians. They would have certainly solved the problem of Menecno without resorting to conics. We consider it absolutely noteworthy that the solution of the second degree equations was known to the Babylonians, who had made it a working tool[6]. Probably the Babylonians had come to understand that the solution of a second degree equation is reducible to the extraction of a square root and to the solution of a first degree equation. We will discuss the so-called Babylonian algorithm for calculating square roots in the next chapter.

In order to better appreciate the Babylonian method, let us remember that the logical steps that lead to the solution of one second degree equation are the following:

1. Factorize the second degree polynomial as

$$(ax^2+bx+c)=a\left[x^2+\frac{b}{a}x+\frac{c}{a}\right]=a\left[\left(x+\frac{b}{2a}\right)^2-\left(\frac{b^2-4ac}{4a^2}\right)\right];$$
$$(1.4.1)$$

2. Keep the square root and determine the two distinct roots of the equation

$$\left(x+\frac{b}{2a}\right)^2=\left(\frac{b^2-4ac}{4a^2}\right)\Rightarrow \begin{aligned}x_+&=\frac{-b+\sqrt{\Delta}}{2a}\\x_-&=\frac{-b-\sqrt{\Delta}}{2a}\end{aligned}, \quad \Delta=\frac{b^2-4ac}{4a^2}.$$
$$(1.4.2)$$

The example which we report below, and found on a tablet dating back to 4000 BC, is presumably an exercise proposed for educational

---

[6]This aspect will be commented further in this book.

purposes. It represents an extremely important example to understand the path of evolution of Mathematics.

The problem is formulated as follows:

*I added 7 times the side of my square to 11 times its surface and
I got 6.25, how much is the side?*

Translated in a modern mathematical language the solution of
the exercise is provided by the roots of the second degree algebraic
equation

$$11x^2 + 7x = 6.25. \tag{1.4.3}$$

The tablet does not contain a general formula for the solution of
second degree equations but a series of instructions reported below.
For an easy comparison with the "modern rule", bear in mind that

$$a = 11, \qquad\qquad b = 7, \qquad\qquad c = -6.25. \tag{1.4.4}$$

The solution recipe follows the steps listed below.

1. Multiply 11 and 6.25: $(-a \cdot c = 68.75)$;

2. Divide 7 by 2: $\left(\dfrac{b}{2} = 3.5 \right)$;

3. Square it: $\left(\dfrac{b^2}{4} = 12.25 \right)$;

4. Add it to the result in step 1:
$$\left(12.25 + 68.75 = 81 \Rightarrow \frac{b^2 - 4ac}{4} = 81 \right);$$

5. Take the square root: $\left(\sqrt{81} = 9 = \sqrt{\dfrac{b^2 - 4ac}{4}} \right)$;

6. Subtract the result in 2 from 9:
$$\left(9 - 3.5 = 5.5 \Rightarrow \sqrt{\frac{b^2 - 4ac}{4}} - \frac{b}{2} = 5.5 \right);$$

The cooking recipe ends up with the crucial point:

7. Find the number which multiplied by 11 returns 5.5, it will
   provide the length of the square side (this instruction amounts
   to

$$11 \cdot 0.5 = 5.5 \Longleftarrow \frac{1}{a} \left(\sqrt{\frac{b^2 - 4ac}{4}} - \frac{b}{2} \right) \Longleftarrow x_+ = \frac{-b + \sqrt{\Delta}}{2a}$$

(1.4.5)

which is the positive root of the second degree equation).

That's the tool kit!

No explanation why it goes that way, no attempt to frame it into a
wider context. We have noticed that only one root has been reported.
Although natural to discard the negative root for the solution of a
geometric problem, it seems however that it was systematic, even
when two positive solutions were admitted. This is supposedly due
to the fact that those ancient mathematicians did not know that,
in the process of extracting the square root, one can have positive
and negative numbers as well. Apart from these details, it is truly
amazing that a problem, posed (and solved in its essential lines) 5000
years ahead of the era of the great algebraic masters of the modern
era, has been "forgotten" for a few millennia.

Babylonians were also able to calculate cubic roots. They might
have been able to solve quite straightforwardly the problem of cube
duplication. They had "tables" of exact cubic roots, if the number
from which to extract the root had not been listed they used an
algorithm, which from our (modern) perspective is reduced to the
identity

$$\sqrt[3]{a} = \sqrt[3]{\frac{a}{b}} \cdot \sqrt[3]{b}$$

(1.4.6)

where b is chosen to be an exact cube root.

The example we give is taken from a tablet in which it was required to evaluate the cubic root of 729000 that is written as

$$\sqrt[3]{729000} = \sqrt[3]{\frac{729000}{27000}} \cdot \sqrt[3]{27000} = \sqrt[3]{27} \cdot 30 = 90$$

(1.4.7)

The scribe, presenting the exercise, took care to choose known roots, if they were not, the procedure to be followed is a sort of interpolation, as we will see in the next chapters.

Let us now summarize what has been discussed so far. We gave a bird's eye view, in perfect Assyrian-Babylonian style, over four millennia of Mathematics. In our "tour" we mixed ancient and modern, without taking care of the logical connection and resorting to concepts (eventually true) used, for our purposes, in "extemporaneous" form, that is, not inserted in a coherent "corpus", such as the canonized one in the Elements of Euclid.

1.5   Numbers, Basis and Polynomials

One of the highest accomplishments of Mathematics has been the discovery of the decimal numeral system which is the way suggested by Arabs and Indus to express numbers in the current notation. Any number can be ordered in powers of 10, accordingly the number 231 is written as

$$231 = 2 \cdot 10^2 + 3 \cdot 10 + 1 \cdot 10^0 = \{2, 3, 1\}_{10}. \qquad (1.5.1)$$

The notation we have foreseen consists of a series of numbers (the composing digits) in brace brackets and of a further number (the base), appended as lower index to the right bracket. The further rule is that the inner bracket numbers are non negative integers less than 10 (all digits $0, 1, \ldots, 9$). According to these prescriptions the base is not unique (10 was perhaps chosen in correspondence of the number of human fingers) but any other positive integer may be employed as well. The use of the base 7 yields

$$\{2, 3, 1\}_7 = 2 \cdot 7^2 + 3 \cdot 7 + 1 \cdot 7^0 = 120. \qquad (1.5.2)$$

The use of a more generic p and an inner bracket triad a, b, c arranged as

$$\{a, b, c\}_p = ap^2 + bp + c \qquad (1.5.3)$$

represents another way of expressing a second degree polynomial. The use of this notation suggests that the numbers $(a^2, 2a, 1)_p$ are all perfect squares, we have indeed

$$\{a^2, 2a, 1\}_p = a^2p^2 + 2ap + 1 = (ap + 1)^2. \tag{1.5.4}$$

The examples we give below helps to better understand the meaning of the notation

$$\{1, 2, 1\}_3 = 16_{10} = \left(4^2\right)_{10},$$
$$\{1, 2, 1\}_5 = 36_{10} = \left(6^2\right)_{10}, \tag{1.5.5}$$
$$\{1, 2, 1\}_{10} = 121_{10} = \left(11^2\right)_{10}.$$

If we interpret any number as a polynomial in a given basis, we can handle it according to the current algebraic rules for polynomials. The operation of the reduction of a number in factors can be interpreted as a polynomial decomposition. We get therefore

$$\{a, b, c\}_p = a\,(p - n_+)\,(p - n_-), \qquad n_\pm = \frac{-b \pm \sqrt{b^2 - 4ac}}{2a}. \tag{1.5.6}$$

It is evident that such a factorization is always possible but it may not result into integers and/or primes. Sometimes it happens. The number $(1, 4, 3)_p$ is 143 in base 10, its prime factors are 11 and 13, as it follows from the factorization procedure

$$143 = \{1, 4, 3\}_{10} = [1 \cdot (p + 1)(p + 3)]_{p=10} = 11 \cdot 13. \tag{1.5.7}$$

We have now acquired sufficient independence of thought to access a first initiation of the Vedic mysteries. We will provide a few examples showing how some products can be worked out, according to the prescriptions of Vedic Math. The method is named after the proponent or the "school" he belonged to, we are not able to transliterate from the Sanskrit and it is not reported here.

We consider the product of two numbers n and m composed by two digits in base 10 and perform the product according to the following rule in which the symbol | is used to indicate by two sets of digits (before and after it).

$$n \times m = \frac{n}{m} \left| \begin{array}{c} n - p \\ m - p \end{array} \right. = n + (m - p)|(n - p)(m - p) \tag{1.5.8}$$

where p is a suitably chosen integer. The technicalities involved in the previous identity are summarized by the mnemonic table in Fig. 1.9 which should be interpreted as follows. The product of two numbers

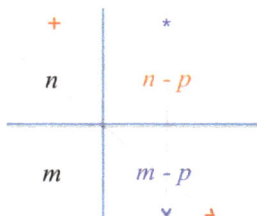

Figure 1.9: $n+(m-p) \mid (n-p)\cdot(m-p)$ ...

$m \cdot n$ is $n + m - p|(n - p)(m - p)$, namely a number composed of two parts. The digits before | are the sum of the numbers n and m decreased by p, those after | are the result of the product of the differences between n, m numbers minus p. Things are easier doing than saying, so choosing $p = 10$ we multiply 13 and 12, according to the previous prescription, as

$$13 \times 12 = \frac{13}{12} \bigg| \frac{13 - 10}{12 - 10} = \frac{13}{12} \bigg| \frac{3}{2} = 13 + 2|3 \times 2 = 15|6 = 156.$$

(1.5.9)

The choice of p is just matter of convenience. By keeping $p = 100$ we find for example

$$111 \times 109 = \frac{111}{109} \bigg| \frac{111 - 100}{109 - 100} = \frac{111}{109} \bigg| \frac{11}{9} = 111 + 9|11 \times 9$$
$$= 120|99 = 12099.$$

(1.5.10)

The number p is not assumed to be smaller than m or n, we can for example multiply 91 by 96 by choosing $p = 100$ as shown below

$$91 \times 96 = \frac{91}{96} \bigg| \frac{91 - 100}{96 - 100} = \frac{91}{96} \bigg| \frac{-9}{-4} = 91 + (-4)|(-9)(-4)$$
$$= 87|36 = 8736.$$

(1.5.11)

The presence of negative numbers is not a big deal for the procedure. The rule of signs yields positive numbers. The same does not happen if p is larger than m or n but not of both, in this case we should proceed according to this slight amendment

$$108 \times 97 = \frac{108}{97} \begin{array}{|c} 108 - 100 \\ \hline 97 - 100 \end{array} = \frac{108}{97} \begin{array}{|c} 8 \\ \hline -3 \end{array} = 108 + (-3)|8 \cdot (-3)$$

$$= 105|(-24) = 105 \cdot 100 - 24 = 10476.$$

$$(1.5.12)$$

Some pitfalls hidden in the procedure must be avoided, we consider the product $21 \cdot 11$ which should return a number of three digits. If we choose $p = 10$, the rule yields

$$21 \times 11 = \frac{21}{11} \begin{array}{|c} 11 \\ \hline 1 \end{array} = 22|11 = (22 + 1)|1 = 231. \qquad (1.5.13)$$

If we apply the Vedic rule as illustrated so far we get a four digit number and this cannot happen. The way out is just to keep the first digit of the number on the right and add it to that on the left (see the part in red of the previous formula). The same rule holds for the product $115 * 123$, which after choosing $p = 100$ is evaluated as

$$115 \times 123 = \frac{115}{123} \begin{array}{|c} 15 \\ \hline 23 \end{array} = 138|345 = (138 + 3)|45 = 14145. \quad (1.5.14)$$

The operation of squaring a number follows the same pattern, we get indeed

$$94 \times 94 = \frac{94}{94} \begin{array}{|c} -6 \\ \hline -6 \end{array} = 88|36 = 8836. \qquad (1.5.15)$$

It is not difficult to realize that the practical rules, we have just described, find their justification in the polynomial nature of numbers and in the algebraic decomposition method we discussed earlier (the proof is straightforward and left as an exercise). The method becomes a useful tool after an adequate training, we therefore suggest the following exercises.

Exercise 1.

1) $97 \cdot 80$,	2) $87 \cdot 89$,	3) $87 \cdot 98$,
4) $87 \cdot 95$,	5) 95^2,	6) $79 \cdot 96$,
7) $98 \cdot 96$,	8) $92 \cdot 99$,	9) 88^2,
10) $97 \cdot 56$,	11) $97 \cdot 63$,	12) $92 \cdot 96$.

The choice of p requires care to avoid problems. Regarding for example the first product the choice of $p = 80$ may create apparent troubles, we invite the reader to explain why.

Let us break the rules we have stipulated, by inserting a negative digit in the polynomial definition of a number in base p, namely

$$\{a, b_-, c\}_p = ap^2 - bp + c, \tag{1.5.16}$$

in other words, pushing further the analogy with polynomials, we have just inserted a negative coefficient, in the definition of a number in base p. We can take advantage on this notation to write

$$\{1, 2_-, 1\}_{10} = 10^2 - 2 \cdot 10 + 1 = (10 - 1)^2 = 9^2,$$
$$\{1, 2_-, 1\}_5 = 4^2, \tag{1.5.17}$$
$$\{1, 2_-, 1\}_3 = 2^2.$$

It is instructive to show that

$$\{1, 2_-, 1\}_{13} = 12^2 \tag{1.5.18}$$

and more in general that

$$\{1, 2_-, 1\}_{1+N} = N^2, \qquad \{1, 2, 1\}_{1-N} = N^2. \tag{1.5.19}$$

Before concluding this paragraph we could try to anticipate what we will discuss in the next chapters, or the so-called divisibility criteria of a given number for a prime. Let us therefore consider the number

$$N(p) = \{a_n, a_{n-1}, .., a_0\}_p = \sum_{s=0}^{n} a_s p^s \tag{1.5.20}$$

and ask ourselves if it can be divided by $p + 1$. According to the divisibility criteria of polynomials, the following condition $N(-1) = 0$ must be satisfied, which amounts to writing

$$0 = \sum_{s=0}^{n} (-1)^s a_s. \tag{1.5.21}$$

The reader may apply this rule to check that the number 132 can be divided by 11 and that $(1, 3, 2)_4$ can be divided by 5.

In this introductory chapter we just gave a flavor of what we will treat in the sequel, where we will treat in depth what we have just mentioned thus far.

Chapter 2

Divisibility Criteria, Osculator Numbers and Roots

2.1 Introduction

At the end of the previous chapter we pointed out that, from a conceptual point of view, the search for a divisibility criterion of one number for another is not unlike the process leading to search the factorization of a polynomial or, which is the same, to find the roots of an algebraic equation. Albeit the statement is correct, in line of the principle, its applicability is very much limited.

Sometimes the procedure may be successful, for example regarding the number $\{1,4,3\}_p$ we find that it corresponds to $(p+1)(p+3)$ which amounts to the factorization in primes for the number $143 = 11 \cdot 13$ (for the base 10). On the other side, for the number $\{1,2,2\}_p$ the procedure does not yield a factorization in the field[1] of real, albeit for $p = 10$ it is $112 = 7 \cdot 2^4$. Finally the number $\{1,5,2\}_p$ admits a decomposition in terms of irrational numbers.

The rules of the number factorization are clear, the reduction must

---

[1] A field in Mathematics is a set with two operations defined on it.

lead to products made up of positive integers and, in the case of decomposition into prime factors, a further restriction is implicit in the request that those integers must be prime numbers[2].

The caveats were necessary to avoid elementary traps. Notwithstanding let us collect a few other consequences of treating a number as a polynomial $N(p)$.

We have already recognized that the criterion of divisibility by 11 (for $p = 10$) can be summarized by the condition that $N(-1) = 0$. We can also convey that, for a generic base p, the same condition ensures that the number can be divided by $p+1$. This statement can be recognized to be a consequence of the *Remainder Theorem*, according to which a polynomial $P(x)$ can be divided by the binomial $x + x_0$ only if $P(-x_0) = 0$. However, regarding integer numbers, we can also conclude that if $d = p + p_0$ and the remainder $N(-p_0)$ is a multiple of d, then the number is divisible by d^3.

Let us therefore consider the number

$$N(p) = \{a_n, \ldots, a_0\}_p \qquad (2.1.1)$$

and take $p = 10$ ($p_0 = -3$). We can conclude that the number is divisible by 13 if

$$N(-3) = \sum_{s=0}^{n}(-3)^s a_s. \qquad (2.1.2)$$

A few further examples are given below

$$52 = \{5, 2\}_{10} \quad \rightarrow \quad N(-3) = -3 \cdot 5 + 2 = -13 \qquad \rightarrow Divisible,$$
$$377 = \{3, 7, 7\}_{10} \quad \rightarrow \quad N(-3) = 3^2 \cdot 3 - 3 \cdot 7 + 7 = 13 \quad \rightarrow Divisible,$$
$$169 = \{1, 6, 9\}_{10} \quad \rightarrow \quad N(-3) = 3^2 \cdot 1 - 3 \cdot 6 + 9 = 0 \quad \rightarrow Divisible,$$
$$439 = \{4, 3, 9\}_{10} \quad \rightarrow \quad N(-3) = 3^2 \cdot 4 - 3 \cdot 3 + 9 = 36 \quad \rightarrow Not\ Divisible.$$
$$(2.1.3)$$

---

[2]For a gentle introduction to the subject see
www.mathsisfun.com/numbers/prime-factorization-tool.html.

[3]If we wanted to use a modern and rigorous language, the previous statement should be written as $\frac{N(p)}{(p+p_0)} = k(p + p_0)$, $\forall k \in \mathbb{Z}$, where $\forall k \in \mathbb{Z}$ means for any number k belonging to integers commutative ring $\mathbb{Z}$.

The same numbers in base $p = 8$ $(p_0 = 5)$ writes

$$52 \; = \{6, 4\}_8 \;\;\; \rightarrow \; N(-5) = -5 \cdot 6 + 4 = -26 \;\;\;\;\;\;\;\; \rightarrow Divisible,$$
$$377 = \{5, 7, 1\}_8 \; \rightarrow \; N(-5) = 5^2 \cdot 5 - 5 \cdot 7 + 1 = 91 \;\; \rightarrow Divisible,$$
$$169 = \{2, 5, 1\}_8 \; \rightarrow \; N(-5) = 5^2 \cdot 2 - 5 \cdot 5 + 1 = 26 \;\;\; \rightarrow Divisible,$$
$$439 = \{6, 6, 7\}_8 \; \rightarrow \; N(-5) = 5^2 \cdot 6 - 5 \cdot 6 + 7 = 127 \rightarrow Not \; Divisible.$$
$$(2.1.4)$$

We have established the criteria of divisibility by the use of a point of view requiring the knowledge of the properties of the polynomials and of some properties of the algebraic equations. The problem can still be treated through the use of more elementary methods. They can sometimes be more "practical" (but not always), so we will review them in following section by comparing them, with those deriving from the polynomial perspective and then with those adopted in Vedic math.

2.2 Elementary Divisibility Criteria

The divisibility criteria in decimal number system for lowest primes belong to math alphabetization. Any school pupil knows that

a) **Divisibility by 3**
 A number can be divided by 3 if the sum of the composing digits is a multiple of 3;

b) **Divisibility by 5**
 A number can be divided by 5 if the number ends by 0 or 5;

c) **Divisibility by 7**
 The divisibility by 7 will be illustrated through an example by checking that 315 is such. The procedure to be followed is given below

 i) Get the "reduced" number 31 by eliminating the last digit (5);

 ii) Subtract to it the double of the eliminated digit $(31 - 2 \cdot 5 = 21)$;

iii) If the resulting number is a multiple of 7 then the number is divisible by 7.

Being 21 a multiple of 7, 315 can be divided by 7.

We can apply the same procedure to check whether 3388 divided by 7 yields an integer. The procedure follows the steps mentioned before

$$i)\ 338 \to 33; \qquad\qquad ii)\ 338 - 8 \cdot 2 = 322. \qquad (2.2.1)$$

We apparently have a problem, because we do not know if 322 is a multiple of 7. In this case we have just to iterate the same steps as before, namely

$$i)\ 322 \to 32; \quad ii)\ 32 - 2 \cdot 2 = 28\ \Rightarrow\ Divisible. \qquad (2.2.2)$$

The rule of the division by 7 is summarized below for convenience.

A number can be divided by 7 if the reduced number, obtained by excluding from it the digit of unit, subtracted by the double of the last digit is a multiple of 7.

We have already mentioned that the previous prescriptions hold for the decimal system. The number $\{1, 2, 3\}_p$ is easily checked to be not divisible by 3 for $p = 5$ and $\{1, 3, 5\}_p$ is not divisible by 5 in base 7. The same holds for the divisibility criterion by 7 and, regarding this point, we note that in base 14 any number is divided by 7 if it ends by 7 or by 0.

Exercise 2. *A nice exercise we propose to the reader is to get the divisibility criteria for 3 and 5 in different basis (say $p = 7, 11 \dots$).*

We report two further criteria regarding the divisibility of a number (in base 10) by 13 and 17. The rule goes as follows.

d) Divisibility by 13
A number can be divided by 13 if the associated reduced number increased by 4 times the discarded digit is a multiple of 13.

The relevant example is given below for the number 169.

i) $169 \to 16$, ii) $16 + 4 \cdot 9 = 52 \Rightarrow$ *Divisible*; (2.2.3)

e) Divisibility by 17
A number can be divided by 17 if the associated reduced number increased by 12 times the discarded digit is a multiple of 17.

The direct application of the above rule allows to check 11067 can be divided by 17. A few words are sufficient to explain why the previous rules work.

Let us therefore consider a three digit number $\{a, b, c\}_{10}$ and the associated reduced form $\{a, b\}_{10}$. According to the prescription for the divisibility by 7, we can set

$$\{a, b\}_{10} - 2c = k \cdot 7, \qquad (2.2.4)$$

multiplying both sides by 10 we find

$$\{a, b, c\}_{10} - 21c = (10k)7, \qquad (2.2.5)$$

which, after factorizing 21 in prime numbers, yields

$$\{a, b, c\}_{10} = (10k + 3c)7 \qquad (2.2.6)$$

and this identity ensures the divisibility of the number by 7. The extension to an arbitrary number of digits is straightforward and is therefore omitted.

The same procedure applied to the divisibility by 13 and 17 yields respectively

$$\{a, b, c\}_{10} = (10k - 3c)13 ,$$
$$\{a, b, c\}_{10} = (10k - 7c)17. \qquad (2.2.7)$$

It is not difficult to convey that the divisibility criterion by 19 and 23 do fit the procedure and we find

$$\{a, b, c\}_{10} = (10k - c)19 \ ,$$
$$\{a, b, c\}_{10} = (10k - 3c)23.$$

(2.2.8)

The associated rules can be worded as

f) **Divisibility by 19**
   A number can be divided by 19 *if the associated reduced number increased by twice the discarded digit is a multiple of* 19.

g) **Divisibility by 23**
   A number can be divided by 23 *if the associated reduced number increased by* 7 *times the discarded digit is a multiple of* 23.

Before closing the section we like to take notice of the fact that all the previously enounced criteria foresee a multiplier of the unit digit. We accordingly note the correspondence

$$7 \rightarrow -2, \qquad 13 \rightarrow 4, \qquad 17 \rightarrow 12,$$
$$19 \rightarrow 1, \qquad 23 \rightarrow 7.$$

(2.2.9)

Their role will be discussed in a wider context in the forthcoming section.

2.3   Vedic Divisibility Criteria

In the previous section we applied our "modern" knowledge on the properties of polynomials to understand the reasons underlying the recipes to guess whether an integer can be divided by a given prime.

We compare the methods we have just illustrated with those prescribed by the ancient Veda mathematicians in which the concept of *Osculator Number* (*ON*) plays a central role.

1. Define the *ON* (we will say later what it is) associated with the divisor number (*DN*);

2. Determine the reduced number (RN) of the Dividend number (DDN);

3. Add to RN the ON multiplied by the discarded digit;

4. If the obtained number is a multiple of the divisor then $\frac{DDN}{DN}$ is an integer;

5. The process can be iterated if the result from point 4) is not definitive.

Example 1. *Let us apply the procedure by checking the divisibility of the number* 13174584 *by* 23. *We anticipate that the ON of* 23 *is* 7 *(be patient until the next subsection) and proceed as shown below.*

$$i) \; 13174584 \; \rightarrow \; 1317458, \qquad ii) \; 1317458 + 7 \cdot 4 = 1317486. \tag{2.3.1}$$

The number we have obtained is rather huge; we need therefore a further iteration

$$i) \; 1317486 \; \rightarrow \; 131748, \qquad ii) \; 131748 + 7 \cdot 6 = 131790. \tag{2.3.2}$$

The number we have obtained ends by 0, *it will not be taken into account and we proceed by discarding* 9.

$$i) \; 13179 \; \rightarrow \; 1317, \qquad ii) \; 1317 + 7 \cdot 9 = 1380, \tag{2.3.3}$$

$$i) \; 138 \; \rightarrow \; 13, \qquad ii) \; 13 + 7 \cdot 8 = 69. \tag{2.3.4}$$

69 *is a multiple of* 23, *End of the Game!*

We have, so far, scratched on the surface of elementary criteria of divisibility. We have foreseen an "empirical" divisibility criterion based on the use of the osculator (a multiplier like those mentioned above). We will see, in the following paragraphs, how this tool may be exploited to extend our numerical clairvoyance.

2.4 Osculator and Numbers

In geometry the osculating circle approximates a curve at the second order in the neighborhood of the tangency point (see Fig. 2.1). Therefore the radius of the circle coincides with the radius of curvature of the curve at that point.

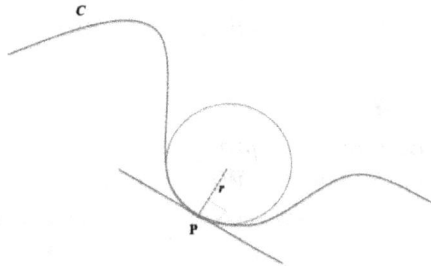

Figure 2.1: Osculating circle.

The origin of the term osculator traces back to the Latin verb "osculari" which means to kiss. The circle and the curve in Fig. 2.1 appear to kiss each other. It should however be noted that

"Thus with an osculator I die." Romeo and Juliet, Act 5, Scene 3.

or

"O, let me osculate that hand!" *"Let me wipe it first; it smells of mortality."* King Lear, Act 4, Scene 6.

Lose their poetical strength.

In Vedic Math there are numbers that are ON of others and, like already seen, they are used to establish the divisibility of one number for another. We will define the method for the computation of an ON of a given number n, indicated with $\Omega(n)$, according to the rules set out below.

1. **ON of number ending by 9**
   Eliminate the last digit (namely 9) and add 1 to what remains

$$\begin{aligned}
\Omega(9) &\rightarrow 0+1=1, \\
\Omega(19) &\rightarrow 1+1=2, \\
\Omega(29) &\rightarrow 2+1=3.
\end{aligned} \qquad (2.4.1)$$

2. **ON of number ending by 7**
   Multiply the number by 7 and apply the procedure in 1).

$$\begin{aligned}
\Omega(7) &\rightarrow 7\cdot 7 = 49 &\rightarrow 4+1 = 5 \\
\Omega(17) &\rightarrow 17\cdot 7 = 119 &\rightarrow 11+1 = 12 \\
\Omega(37) &\rightarrow 37\cdot 7 = 259 &\rightarrow 25+1 = 26.
\end{aligned} \qquad (2.4.2)$$

3. **ON of number ending by 3**
   Multiply the number by 3 and then apply the procedure in 1).

$$\begin{aligned}
\Omega(3) &\rightarrow 3\cdot 3 = 9 &\rightarrow 0+1=1, \\
\Omega(13) &\rightarrow 13\cdot 3 = 39 &\rightarrow 3+1=4, \\
\Omega(23) &\rightarrow 23\cdot 3 = 69 &\rightarrow 6+1=7.
\end{aligned} \qquad (2.4.3)$$

4. **ON of number ending by 1**
   Multiply the number by 3 or by 9 and then apply the previous indications.

$$\begin{aligned}
\Omega(31) &\rightarrow 31\cdot 3 = 93 &\rightarrow 93\cdot 3 = 279 &\rightarrow 27+1=28, \\
\Omega(41) &\rightarrow 41\cdot 3 = 123 &\rightarrow 123\cdot 3 = 369 &\rightarrow 36+1=37, \\
\Omega(51) &\rightarrow 51\cdot 3 = 153 &\rightarrow 153\cdot 3 = 459 &\rightarrow 45+1=46.
\end{aligned}$$
$$(2.4.4)$$

In the forthcoming section we will clarify the reasons underlying the recipes given in the previous points 1-4.

2.5 ON and Divisibility

We will provide here a few examples which corroborate the utility of the ON multiplier.

Example 2. *Use the ON* $\Omega(17) = 12$, *to prove that* 425 *is divisible by* 17.

$$i)\ 425\ \rightarrow\ 42, \qquad\qquad ii)\ 42 + 12 \cdot 5 = 102\ \rightarrow\ 10 + 12 \cdot 2 = 34.$$
$$(2.5.1)$$

The number of iterations increases with the number of digits composing either dividend and *ON* but, never despair, the procedure will bring you to the right answer!

Example 3. *An example is given by the proof of the divisibility by* 17 *of* 83521

$$
\begin{array}{ll}
i)\ 83521\ \rightarrow\ 8352, & ii)\ 8352 + 12 \cdot 1 = 8364, \\
i)\ 8364\ \rightarrow\ 836, & ii)\ 836 + 12 \cdot 4 = 884, \\
i)\ 884\ \rightarrow\ 88, & ii)\ 88 + 12 \cdot 4 = 136, \\
i)\ 136\ \rightarrow\ 13, & ii)\ 13 + 12 \cdot 6 = 85, \\
i)\ 85\ \rightarrow 8, & ii)\ 8 + 12 \cdot 5 = 68, \qquad (2.5.2) \\
i)\ 68\ \rightarrow 6, & ii)\ 6 + 12 \cdot 8 = 102, \\
i)\ 102\ \rightarrow 10, & ii)\ 10 + 12 \cdot 2 = 34, \\
i)\ 34\ \rightarrow 3, & ii)\ 3 + 12 \cdot 4 = 51, \\
i)\ 51\ \rightarrow 5, & ii)\ 5 + 12 \cdot 1 = 17.
\end{array}
$$

A cursory look to the prescriptions in eqs. (2.4.4)-(2.4.1) reveals that there is a one to one correspondence between the *ON* and the multiplier of the ordinary divisibility criteria. All but 7 for which $\Omega(7) = 5$. We use it to check the divisibility by 7 of 2072.

Example 4.

$$
\begin{array}{ll}
i)\ 2072\ \rightarrow\ 207, & ii)\ 207 + 5 \cdot 2 = 217, \\
i)\ 217\ \rightarrow\ 21, & ii)\ 21 + 5 \cdot 7 = 56, \qquad (2.5.3) \\
i)\ 56\ \rightarrow 5, & ii)\ 5 + 5 \cdot 6 = 35.
\end{array}
$$

We have accordingly found that it does. We may also think that 35 *is not sufficient to close the game and we like to achieve a number closer to 7, we just iterate the procedure and find*

$$
\begin{array}{ll}
i)\ 35\ \rightarrow\ 3, & ii)\ 3 + 5 \cdot 5 = 28, \\
i)\ 28\ \rightarrow\ 2, & ii)\ 2 + 5 \cdot 8 = 42, \\
i)\ 42\ \rightarrow\ 4, & ii)\ 4 + 5 \cdot 2 = 14, \qquad (2.5.4)\\
i)\ 14\ \rightarrow\ 1, & ii)\ 1 + 5 \cdot 4 = 21, \\
i)\ 21\ \rightarrow\ 2, & ii)\ 2 + 5 \cdot 1 = 7.
\end{array}
$$

The described procedure has allowed us to establish a general divisibility criterion through the somewhat obscure concept of the ON. At this point we should reconcile the "Vedic" concepts with those more familiar with "western" arithmetic, outlined in the introductory sections of this chapter.

2.6 Unveiling the Osculator Mistery

In general terms the divisibility of a number $\{a, b, c\}_{10}$ by an integer n is expressed by the condition

$$
\{a, b, c\}_{10} = 10\, kn - (10\, \Omega(n) + 1)c, \qquad k \in \mathbb{Z}. \qquad (2.6.1)
$$

The osculator $\Omega(n)$ is then specified by the further condition $10\, \Omega(n) + 1 = rn$ ($r \in \mathbb{Z}$). We accordingly find

$$
\Omega(n) = \frac{rn + 1}{10}. \qquad (2.6.2)
$$

Requiring furthermore that $\Omega(n)$ be a positive integer we find

$$
\Omega(13) = \frac{3 \cdot 13 + 1}{10} = 4, \qquad\qquad \Omega(17) = \frac{7 \cdot 17 + 1}{10} = 12,
$$

$$
\Omega(19) = \frac{1 \cdot 19 + 1}{10} = 2, \qquad\qquad \Omega(23) = \frac{3 \cdot 23 + 1}{10} = 7.
$$

$$
(2.6.3)
$$

What about $\Omega(7)$? If we do relax the condition that r and $\Omega(.) \in \mathbb{N}$, we find, by choosing $r = -3$

$$
\Omega(7) = \frac{-3 \cdot 7 + 1}{10} = -2, \qquad (2.6.4)
$$

according to the rules of Section 2.2. On the other side, by choosing r and $\Omega(.) \in \mathbb{N}^4$, we get

$$\Omega(7) = \frac{7 \cdot 7 + 1}{10} = 5. \tag{2.6.5}$$

It is evident that, accepting negative integer osculators, we find, for the divisors 19 and 23, the alternative values

$$\Omega(19) = \frac{-9 \cdot 19 + 1}{10} = -17 \tag{2.6.6}$$

and

$$\Omega(23) = \frac{-7 \cdot 23 + 1}{10} = -16. \tag{2.6.7}$$

It should also be underscored that the osculator is not unique and, for example

$$\Omega(7) = \frac{77 \cdot 7 + 1}{10} = 54, \qquad \qquad \Omega(7) = \frac{-33 \cdot 7 + 1}{10} = -23.$$
$$\tag{2.6.8}$$

The statement is correct but the choice of an osculator larger (in absolute value) than the number it osculates, may result inefficiently from the computational point of view. We add therefore the further condition (whenever possible[5]) $|\Omega(n)| < n$.

The prescriptions given so far holds for numbers in base 10. We discuss the extension of the method to an arbitrary basis later in this chapter.

As a matter of fact the examples we have just illustrated yield an idea on how ancient and modern concepts, regarding the elementary divisibility criteria, can be merged within a "coherent" framework.

---

[4]We remind that $\mathbb{N}$, $\mathbb{Z}$, $\mathbb{Q}$, $\mathbb{R}$, $\mathbb{C}$ are the mathematical symbols used to indicate Natural, Integer, Rational, Real and Complex numbers respectively. We will use the form with $.^{\pm}$ for positive or negative numbers and $._0$ to rule out the zero.

[5]We have added "whenever possible" because we do not know whether exist numbers whose osculator is always larger than the number itself, the Vedic rule of Section 2.4 excludes such a possibility, but it is just an empirical recipe. The reader has all the elements to provide a general statement on this point.

Before closing this section we invite the reader to reconsider the criterion of divisibility by 3 according to the prescription of Section 2.2 and prove that it is correct.

Consider the base 10

$$N = \sum_{s=0}^{n} a_s 10^s. \tag{2.6.9}$$

The divisibility criterion impose that

$$\sum_{s=0}^{n} a_s = 3m \tag{2.6.10}$$

which yields

$$a_0 = 3m - \sum_{s=1}^{n} a_s. \tag{2.6.11}$$

The number can therefore be written as

$$N = \sum_{s=1}^{n} a_s 10^s + 3m - \sum_{s=1}^{n} a_s = \sum_{s=1}^{n} a_s (10^s - 1) + 3m, \tag{2.6.12}$$

being $10^s - 1$ divisible by 3 the validity of the criterion is proven (although obvious, we avoid any misunderstanding by noting that $10^2 - 1 = 99$, $10^3 - 1 = 999$, $10^4 - 1 = 9999$).

Consider the case of a generic base p and prove the existence of a more general divisibility criterion along the line of the previous discussion. Write a generic number as

$$N = \sum_{s=0}^{n} a_s p^s \tag{2.6.13}$$

and assume that

$$\sum_{s=0}^{n} a_s = Km \tag{2.6.14}$$

where K and m are integers. By applying the same procedure as before we find

$$N = \sum_{s=1}^{n} a_s(p^s - 1) + Km. \tag{2.6.15}$$

It is evident that if $p^s - 1$ is a multiple of K, the number is divisible by K. If $p = 18$ the number is therefore divisible by 17.

Exercise 3. *Use this criterion to show that the number* 566712 *is divisible by* 17.

(Hint: write the number in base 18.)

Finally state that the divisibility criterion by 5 is so obvious that the relevant proof can be omitted.

2.7 Fermat's Little Theorem, Further Examples and Different Notation

The matter we are exploring may appear very much boring and without any hope of practical utility. This is not true.

The divisibility criteria have significant repercussions in the field of information security. The associated issues are widely studied and various practical methods have been proposed. Within this respect, a significant role has been played by the so-called "Fermat's Little Theorem" FLT which states the following:

Theorem 1 (Fermat's Little Theorem). *If p is prime and a is not divisible by p, then the number $a^p - a = mp$.*

In modern terms, namely by the use of modular arithmetic, the FLT is formulated as follows:

Theorem 2. *If p is prime and a is not a multiple of p, then we get $a^p \equiv a \pmod{p}$ or $a^{p-1} \equiv 1$.*

The notation $s \equiv q \pmod{l}$, which reads "s congruent q modulo l", is an educated way to say that $\frac{s-q}{l} \in \mathbb{Z}$.

Even though it is an off topic, we like to spend a few words about the meaning of the notation.

Example 5. *Suppose we consider the number* 17 *and ask how it should be written in modular arithmetic if we consider* $p = 3$. *The algorithm is the following:*

    a) Consider the closest integer multiple of 3 *less* 17 *(namely* 15*),*
    b) subtract it to 17 *(*$17 - 15 = 2$*),*
    c) write $17 \equiv 2 \pmod{3}$.

It should be underscored that

$$m = m \pmod{k}, \qquad \forall k > m \qquad (2.7.1)$$

and it can therefore be checked, e.g, that

$$6^{3-1} \equiv 2 \pmod{3}, \quad 8^{3-1} \equiv 0 \pmod{3} \qquad (2.7.2)$$

and also the following modular arithmetic table for $\pmod{7}^6$ *which should be read as follows: keeping* 3 *in the vertical column and* 5 *in the horizontal part, we obtain the remainder* 1.

The *FLT* sounds of fundamental importance and indeed it is. It allows the so-called *"test of primality"* namely to check whether a given number is prime or not.

To get a feeling whether a number is prime, it is "sufficient" to choose randomly two numbers (not multiple of p) and check the conditions of the Fermat Theorem. The number 5 is prime and we get

$$3^4 - 1 \equiv 0 \pmod{5}, \qquad 2^4 - 1 = 15 \equiv 0 \pmod{5}. \qquad (2.7.3)$$

---

[6] See for example James Robinson, Fermat's Little Theorem, elaboration on history of Fermat's Theorem and implications of Euler's generalizations by means of the Totient Theorem. Master Thesis, Graduated College University of Arizona, 2011.

Table 2.1: **Mod 7 Arithmetic Table**

	1	2	3	4	5	6
1	1	2	3	4	5	6
2	2	4	6	1	3	5
3	3	6	2	5	1	4
4	4	1	5	2	6	3
5	5	3	1	6	4	2
6	6	5	4	3	2	1

Are we able to prove the *FLT*? Albeit, as usual, Fermat did not give any proof of his theorem. The Swiss mathematician Euler provided a fairly simple demonstration, which we will further simplify using a specific example, which however captures the essence of the rigorous procedure[7]:

Consider $a = 2$, $p = 5$, consider the first $p - 1$ multiples of a, namely $m_1 = i \cdot a$, $i = 1, \ldots, p - 1$ and note that, for our specific case,

$$m_1 \equiv 2 \ (\text{mod } 5), \quad m_2 \equiv 4 \ (\text{mod } 5),$$
$$m_3 \equiv 1 \ (\text{mod } 5), \quad m_4 \equiv 3 \ (\text{mod } 5). \tag{2.7.4}$$

On the other side by multiplying all m_i we find

$$a(2a)(3a) \cdots \cdots ((p - 1)a) = (p - 1)!a^{p-1} \tag{2.7.5}$$

and also, by virtue of (2.7.4),

$$m_1 m_2 m_3 \ldots m_{p-1} = (p - 1)! \ (\text{mod } p). \tag{2.7.6}$$

By comparing (2.7.5) and (2.7.6) we eventually end up with

$$(p - 1)!a^{p-1} \equiv (p - 1)! \equiv (p - 1)! \ (\text{mod } p) \tag{2.7.7}$$

---

[7]See for example A. Bogomolny, 2000, Fermat's Little Theroem, in Interactive Mathematics Miscellany and Puzzles, retrieved on June 11, 2010 from www.cut-theknot.org/blue/Fermat.shtml and C. Caldwell, 1994, Proof of Fermat's Little Theorem, in Prime Pages, retrieved on June 13, 2010, from http://primes.utm.edu/notes/proofs/FermatsLittleTheorem.html.

which turns into the *FLT*.

If you do not catch the argument the first time, take a breath and read again.

The inverse of *FLT*, namely if $a^{p-1} \equiv 1 \pmod{p}$, then p is prime, is not true. Notwithstanding the numbers satisfying this property are of noticeable interest. They are called *pseudo-primes* and share indeed a property common to all primes, without being such.

This is a fascinating topic but, unfortunately we cannot enter the subject, whose "practical" importance is certified by the relevant use in e-commerce to protect the data exchanged via internet.

After this touch on a modern view to divisibility criteria and its possible applications, we go back to the mainstream of our discussion by adding a few comments on the notation. It is not always uniform and this may be source of confusion for beginners. For example the symbol

$$a : n \, ? \tag{2.7.8}$$

is equivalent to the question whether a is divisible by n. The mixing up of divisibility criteria in Sec. 2.2, our notation in (2.7.8) and the previous one for the osculator procedure yields

$$56670{:}7 \, ? \;\rightarrow 5667{:}7 \, ? \;\rightarrow 566 - 14{:}7 \, ? \;\rightarrow 55 - 4 = 51{:}7 \, ?$$
$$\rightarrow Not \; Divisible, \tag{2.7.9}$$

where we have adopted the algorithm[8] with *ON* $\Omega(7) = -2$ The same notation for $\Omega(7) = 5$ yields

$$56670 : 7 \, ? \;\rightarrow 5667 : 7 \, ? \;\rightarrow 601 = 7 \, ? \;\rightarrow\; 65 : 7 \, ? \rightarrow Not \; Divisible. \tag{2.7.10}$$

The notation

$$N : (10p + 1) \tag{2.7.11}$$

----

[8]The term "algorithm" defines a calculation protocol based on a finite number of steps.

is relevant to the divisibility of a number N by $10p + 1$.

The method ensuring the relevant proof is based on the use of the integer part of a number (like, $\lfloor \frac{10}{3} \rfloor = 3$, $\lfloor \frac{5}{2} \rfloor = 2$, $\lfloor \frac{7}{5} \rfloor = 1 \ldots$). In general terms $\lfloor \frac{m}{n} \rfloor$ is an integer less or equal to $\frac{m}{n}$. It is evident that according to this notation the number N can always be cast as

$$N = 10\lfloor \frac{N}{10} \rfloor + u, \tag{2.7.12}$$

where u is the decimal part of $\frac{N}{10}$. The number 2257 can, according to this notation, be written as

$$2257 = 10\lfloor \frac{2257}{10} \rfloor + 7. \tag{2.7.13}$$

Let us now consider a number written as

$$A_1 \cdot 10p + A_0 \tag{2.7.14}$$

where

$$A_1 = \frac{1}{p}\lfloor \frac{N}{10} \rfloor, \qquad\qquad A_0 = u. \tag{2.7.15}$$

The number (2.7.15) is therefore recognized as being composed by two digits in base $10p$. We can therefore generalize the divisibility criterion of a number in base 10 by 11 to state that (2.7.14) is divisible by $10p + 1$ if

$$(-1)\frac{1}{p}\lfloor \frac{N}{10} \rfloor + u = k(10p + 1). \tag{2.7.16}$$

The divisibility is therefore ensured if

$$up - \lfloor \frac{N}{10} \rfloor = kp(10p + 1) \tag{2.7.17}$$

or, by changing sign, if "$\lfloor \frac{N}{10} \rfloor - up$" is an integer multiple of $10p + 1$.

The example given below is helpful to clarify the procedure

$$2257 : 61? \tag{2.7.18}$$

By using the previous prescription we set

$$\left(10\lfloor\frac{2257}{10}\rfloor + 7\right) : (10 \cdot 6 + 1), \qquad (2.7.19)$$

by having established that $u = 7$, $p = 6$, we find

$$\lfloor\frac{N}{10}\rfloor - up = 225 - 7 \cdot 6 = 183 = 3 \cdot 61 \qquad (2.7.20)$$

which confirms the divisibility.

2.8　Squares, Square Roots and All That

In the preceding paragraphs we have seen how the polynomial form of a number can clarify much of the relevant computational "technicalities". It comes therefore out that much of the practical rules of the Vedic Mathematics (and of junior school recipes as well) can be "logically" framed within this context. In the following we will see how this statement applies to the calculus of the powers of a number and to that of square roots.

A perfect square number in a generic base p writes

$$n^2 = (p + 1)^2 = p^2 + 2p + 1 = \{1, 2, 1\}_p = \{1, 2, 1\}_{n-1}, \qquad (2.8.1)$$

for $p = 3$ we get 16 while $p = 10$ yields 121.

A perfect cube can accordingly be written as

$$n^3 = (p + 1)^3 = p^3 + 3p^2 + 3p + 1 = \{1, 3, 3, 1\}_p = \{1, 3, 3, 1\}_{n-1}. \qquad (2.8.2)$$

The following comment helps understanding the meaning of the previous conclusions.

Observation 1. *The rule we have stated, even though trivially true, should be taken with some care. For example the number* 625 *(the square of* 25*) in base* 10 *reads*

$$25^2 = 625 = 6 \cdot 10^2 + 2 \cdot 10 + 5 = \{6, 2, 5\}_{10} \qquad (2.8.3)$$

which is not the form of a perfect square, at least regarding the base 10 polynomial form. It can be recovered by noting that by using $p = 25 - 1$ *we end up with*

$$\{1, 2, 1\}_{24} = 24^2 + 2 \cdot 24 + 1 = 576 + 49 = 625. \qquad (2.8.4)$$

It should be stressed that the method we have envisaged is not practical to get the root of a number, since one has to find the suitable basis p to express the number in the form of a perfect square, cube and so on. The associated algorithmic procedure is more complicated than a less elaborated, but more effective tool, yielding the square or cubic root of a given number. In any case never despair! We can make a step further, and consider squares whose last digit is 5

$$(\{a, 5\}_{10})^2 = a^2 \cdot 10^2 + 2 \cdot 5 \cdot a \cdot 10 + 25 = a(a + 1)10^2 + 25. \ (2.8.5)$$

We can take advantage from the last identity to infer a "general rule" to get the square root of a given number. We start by squaring the number 85. According to the identity (2.8.5), it is found to be a number of four digits, the first two are obtained by multiplying the number 8 by itself increased by 1 ($8 \cdot 9 = 72$), the last two digits are simply 2 and 5, namely

$$85^2 = (\{8, 5\}_{10}) = 8(8 + 1)10^2 + 25 = 72 \mid 25 = 7225. \qquad (2.8.6)$$

Reversing the argument, if we are required to find the square root of 7225, we know that we can proceed by solving the equation $a(a+1) = 72$ to find the first digit and then by appending to it 5.

We can develop a rule to evaluate the square root of a number ending by 25. Regarding 5625 we proceed as reported below:

$$\begin{aligned} 5625 = (\{a, 5\}_{10})^2 &\Rightarrow \sqrt{5625} = \{a, 5\}_{10}, \\ 56 = a(a + 1), \quad a = 7 &\Rightarrow \sqrt{5625} = 7 \mid 7 = 75 \end{aligned} \qquad (2.8.7)$$

We note that the first digit 7 is the (positive solution) of a second degree equation, which admits two solutions:

$$a(a + 1) = m \Rightarrow a_\pm = \frac{-1 \pm \sqrt{1 + 4m}}{2}, \qquad (2.8.8)$$

so what about the negative solution? Denoting with $a_\pm$ the negative and positive solutions, we can write the square root as

$$\{a_+, 5\}_{10} = -\{a_-, 5\}_{10} . \tag{2.8.9}$$

Regarding our specific example, we get

$$75 = -(-8 \cdot 10 + 5) . \tag{2.8.10}$$

We use the rule of the square of numbers ending in 5 to learn how to make products quickly. If we wanted to calculate the product $66 \cdot 64$, we proceed as it follows.

$$(65 + 1)(65 - 1) = 65^2 - 1 = 4225 - 1 = 4224, \tag{2.8.11}$$

where we have used the square difference formula and the rule for the square of numbers ending by 5. The mnemonic application yields for example $74 \cdot 76 = 5624$.

More in general, it is easily argued that the product of two numbers N_+, N_- such that

$$N_+ = N + r, \qquad N_- = N - r, \qquad N_\pm = \{a, 5 \pm r\}_{10} \tag{2.8.12}$$

where N is a two digit number ending by 5 and $r \leq 5$, is

$$N_+ N_- = a(a + 1)10^2 + (25 - r^2). \tag{2.8.13}$$

We find $72 \cdot 78 = 5616$ by using the just outlined rule

$$N_\pm = \{a, 5 \pm r\}_{10} = \{7, 5 \pm 3\}$$
$$\Rightarrow N_+ \cdot N_- = 7(7 + 1)10^2 + (25 - 3^2) = 56 \mid 16 = 5616. \tag{2.8.14}$$

We can now try a simple extension of what we have learnt so far to number written in different basis than 10. We note indeed that if $p = 2k$ then

$$\left(\{a, k\}_p\right)^2 = a(a + 1)p^2 + k^2, \tag{2.8.15}$$

therefore the square of 77, which in base 14 writes $\{5, 7\}_{14}$, is

$$(\{5, 7\}_{14})^2 = 30 \cdot 14^2 + 49 = 5929. \tag{2.8.16}$$

A naïve extension of the previous formula is just given by

$$({\{a, n\}_{2n}})^2 = a(a+1)(2n)^2 + n^2 = {\{a(a+1), n^2\}}_{(2n)^2}. \quad (2.8.17)$$

The rules we have learned are amusing even though useful only after an adequate training. Further comments and more efficient procedures will be discussed in the forthcoming parts of the book.

We can proceed further in the "computational technology" by exploiting the following identity

$$a^2 = (a+b)(a-b) + b^2 \quad (2.8.18)$$

to compute the square of 92, which writes

$$92^2 = (92+8)(92-8)+8^2 = 100\cdot84+64 = 84\cdot10^2+64 = 84 \mid 64 = 8464. \quad (2.8.19)$$

The "rule" we have applied to find out the result is fairly obvious: the square of a number N between 91 and 99 is a 4 digit number, the last two being given by the square of the difference Δ between 100 and the number itself and the first two by $N - \Delta$, namely

$$N^2 = (N - \Delta) \mid \Delta^2. \quad (2.8.20)$$

It is therefore easily stated that

$$93^2 = 86 \mid 49, \qquad 94^2 = 88 \mid 36, \qquad 97^2 = 94 \mid 09, \qquad 98^2 = 96 \mid 04. \quad (2.8.21)$$

We are now ready to make another step forward, by guessing that:

The square root of a number of 4 digits with last two being a perfect square "might" be given by

$$\sqrt{a \mid b \mid c \mid d} = (a \mid b) + \sqrt{c \mid d} \quad (2.8.22)$$

and indeed we find

$$\sqrt{8836} = 88 + \sqrt{36} = 94, \qquad \sqrt{9604} = 96 + \sqrt{04} = 98. \quad (2.8.23)$$

We consider now square of number less than 100 but not close to it

$$87^2 = (87 + 13)(87 - 13) + 13^2 = 100 \cdot 74 + 169$$
$$= 100 \cdot 74 + (100 + 69) = 74 + 1 \mid 69 = 75 \mid 69;$$
$$89^2 = (89 + 11)(89 - 11) + 11^2 = 100 \cdot 78 + 121$$
$$= 100 \cdot 78 + (100 + 21) = 78 + 1 \mid 21 = 79 \mid 21;$$
$$74^2 = (74 + 26)(74 - 26) + 26^2 = 100 \cdot 48 + 676$$
$$= 100 \cdot 48 + (6 \cdot 100 + 76) = 48 + 6 \mid 76 = 54 \mid 76.$$

$$(2.8.24)$$

From the conceptual point of view the previous relations are based on the same criterion we exploited to state the square of numbers like 98. We can therefore establish some tricks to keep the square root of four digit numbers.

We invert the previous procedures and use as example the derivation of the square root of 7569. The "method" we employ consists of the following steps:

1. We separate the first two digits from the remaining part (75 | 69);

2. We determine the digit which, put in front of the last two, yields a perfect square of 3 digits, namely $1 \mid 69 = 169 = 13^2$;

3. Subtract it to the number composed by the two digits and get the square root of our number as $(75 - 1) + 13 = 87$.

Apply the method to the following examples.

Example 6.

$$\sqrt{6889} \rightarrow 68 \mid 89 \rightarrow 68 - 2 \mid 289 \rightarrow 66 \mid 17^2 \rightarrow 66 + 17 = 83;$$
$$\sqrt{6724} \rightarrow 67 \mid 24 \rightarrow 67 - 3 \mid 324 \rightarrow 64 \mid 18^2 \rightarrow 64 + 18 = 82.$$

$$(2.8.25)$$

In this section we have just considered perfect roots, the next part of the chapter will be devoted to a generalization of the methods considered so far and to other complementary practical techniques.

2.9   Vargamula

The ability to compute the square root (Vargamula in Sanskrit) of a number, either by "hand" or by the use of tables, belongs to an ancient epoch. Nowadays every pocket calculator can execute such a task in times certainly faster than those needed by a human operator. Its use is much more comfortable than a bulky handbook of tables and easier to use than the "ruler". Nevertheless it is disappointing that practically no one (except well educated scholars) knows the old methods of calculation and it is too often ignored how the calculator executes the operations necessary for root extraction.

Before remembering old rules, belonging to the western computational "technology", and those (less known) of Vedic recipes book, we outline the fingerprints of a perfect square:

a) The sum of its digits is $1, 4, 7, 9$,

$$49 \rightarrow 13 \rightarrow 4; \quad 121 \rightarrow 4; \quad 6241 \rightarrow 13 \rightarrow 4, \quad 106929 \rightarrow 27 \rightarrow 9;$$
$$(2.9.1)$$

b) The last of its digit is $0, 1, 4, 5, 6, 9$;

c) It can always be reduced to the sum of the first odd numbers[9]

$$N^2 = \sum_{n=0}^{N-1} (2n + 1); \qquad (2.9.2)$$

d) If a number consists of n digits, its square root will be composed by $\frac{n}{2}$ or $\frac{n+1}{2}$ digits.

It is worth noting that conditions *a)* and *b)* are *necessary but not sufficient*. This is an educated expression to state that all perfect squares fulfill such conditions even though other numbers, satisfying

---

[9]For the moment we justify (2.9.2) by checking that

$9 = 1+3+5$, $N = 3$, $16 = 1+3+5+7$, $N = 4$, $25 = 1+3+5+7+9$, $N = 5$, ...

A "proof" is given in Chapter 6.

such a requisite, are not necessarily perfect squares. This is the case of the number 139, which fulfills b) and not a). The number 169 does satisfy both a, b) requirements. The exact root, if any, is composed by two digits, the last of which can safely be assumed to be 3, the first cannot be larger than one, we can therefore guess that the root is 13. It is easily checked that it belongs to squares of type

$$169 = \{1, 6, 9\}_{10}. \tag{2.9.3}$$

Its square root writes

$$\sqrt{\{1, 6, 9\}_{10}} = 10 + 3. \tag{2.9.4}$$

We can accordingly guess that

$$\sqrt{\{1, 6, 9\}_p} = p + 3 \tag{2.9.5}$$

thus getting e.g.

$$\{1, 6, 9\}_{14} = 289, \qquad \sqrt{\{1, 6, 9\}_{14}} = 14 + 3 = 17. \tag{2.9.6}$$

Regarding the "rule" we have drawn to compute square root, let us consider the number 127, which is not a perfect square, and write it in the form

$$127 = \{1, 6, 9\}_{p=\sqrt{127}-3} \tag{2.9.7}$$

which yields

$$\sqrt{\{1, 6, 9\}_{p=\sqrt{127}-3}} = p + 3 = \sqrt{127}. \tag{2.9.8}$$

Not a great discovery!!! But at least a check of the correctness of the method.

The number 1849 (ending by 9 with the sum of its digits $= 4$) is a good candidate for being a perfect square. According to the properties a, b, d we can assume that the root is composed by two digits, that the last might be 3 and that the first not larger than 4. It is not difficult to state that

$$\sqrt{1849} = 4 \cdot 10 + 3 = 43. \tag{2.9.9}$$

It is just the result of a number of guesses, which can be framed within a more general computational protocol, reported below:

i) Draw a table with five rows and six columns and dispose the number (the radicand) in the first row as illustrated in the scheme (2.9.10). The table boxes will be denoted by (row, column);

$$\sqrt{1849} \rightarrow \begin{array}{|c|c|c|c|c|c|} \hline 1 & 8 & 4 & 9 & & \\ \hline 1 & 6 & & & 4 & \\ \hline & 2 & 4 & & 8 & \\ \hline & & 9 & 3 & & \\ \hline & & 0 & & & \\ \hline \end{array} \rightarrow \sqrt{1849} = 43 \qquad (2.9.10)$$

ii) Determine the largest square less or equal to the number composed by the first two digits of the radicand (namely 16) and write it in boxes $(2, 1-2)$ and its root in box $(2, 6)$;

iii) Execute the difference between the first two digits (namely 18 and 16) and allocate it in $(3, 2)$ and next to it (box $(3, 3)$) write the digit in $(1, 3)$;

iv) Keep a digit which is a divisor of the number composed by the digits in boxes $(3, 2 - 3)$, and put it in box (3.5). We choose 8 (see below for further instructions);

v) Divide the number in $(3, 2 - 3)$ by that in $(3, 5)$ and put it in $(4, 6)$;

vi) Calculate its square and put it in $(4, 4)$;

vii) Make the difference between the last digit of the radicand and the number in box $(4, 4)$, if it is *zero*, the radicand is a perfect square having a two digits root composed by the numbers in boxes $(2, 6)$ and $(4, 6)$.

It is however worth stressing that the procedure, at least for the way it has been illustrated, is doubtful and requires some experience. The choice of 8 is a matter of convenience. It has been determined in such a way: it is a divisor of the number with digits $(3, 2 - 3)$ and the square of the quotient is the last digit of the radicand. The following examples clarify what we mean.

Example 7. *If we are interested in finding the square root of* 2809, *we apply an analogous procedure and get*

$$\sqrt{2809} \rightarrow \begin{pmatrix} 2 & 8 & 0 & 9 & - & - \\ 2 & 5 & - & - & - & 5 \\ - & 3 & 0 & - & 10 & - \\ - & - & - & 9 & - & 3 \\ - & - & - & 0 & - & - \end{pmatrix} \rightarrow 53. \qquad (2.9.11)$$

The number we have chosen (reported in red) is a divisor of 30 *with a quotient which squared yields* 9.

Example 8. *A slight modification of the procedure yields for the root of* 2209 *the result*

$$\sqrt{2209} \rightarrow \begin{pmatrix} 2 & 2 & 0 & 9 & - & - \\ 2 & 5 & - & - & - & 5 \\ - & (-3) & 0 & - & 10 & - \\ - & - & - & 9 & - & \textbf{-3} \\ - & - & - & 0 & - & - \end{pmatrix} \rightarrow \{5, 3_-\}_{10} = 50-3 = 47.$$

$$(2.9.12)$$

The solution we have adopted has been that of choosing as first step an exact square larger or equal to the number formed by the two first digits. The price to be paid is that the root is expressed in base 10 *with a negative coefficient.*

Example 9. *For* $\sqrt{3364}$ *we obtain*

$$\sqrt{3364} \rightarrow \begin{pmatrix} 3 & 3 & 6 & 4 & - & - \\ 3 & 6 & - & - & - & 6 \\ - & (-3) & 6 & - & 18 & - \\ - & - & - & 4 & - & (-2) \\ - & - & - & 0 & - & - \end{pmatrix} \rightarrow \{6, 2_-\}_{10} = 58.$$

$$(2.9.13)$$

Example 10. *Finally, we note that*

$$\sqrt{6241} \rightarrow \begin{pmatrix} 6 & 2 & 4 & 1 & - & - \\ 6 & 4 & - & - & - & 8 \\ - & (-2) & 4 & - & 24 & - \\ - & - & - & 1 & - & (-1) \\ - & - & - & 0 & - & - \end{pmatrix} \rightarrow \{8, 1_-\}_{10} = 79.$$

$$(2.9.14)$$

The extension of the method to a generic base p could be a good exercise for the reader.

We have so far dealt with perfect squares but a significant amount of work has been done for approximated methods, one of these known since the antiquity is the so-called *Heron iterative method* (also known as Babylonian algorithm). The method can be illustrated as follows.

If R is the radicand, its square root is obtained through the following iteration

$$r_{k+1} = \frac{1}{2}\left(r_k + \frac{R}{r_k}\right), \qquad k = 0, 1, 1, \ldots \qquad (2.9.15)$$

The starting value r_0 is any value approximating the square root of R, by keeping $R = 30$ and $r_0 = 5$, namely a value less than the effective value $(5.4772255751\ldots)$ we get

$$r_1 = \frac{1}{2}\left(5 + \frac{30}{5}\right) = 5.5, \qquad r_2 = \frac{1}{2}\left(5.5 + \frac{30}{5.5}\right) = 5.47727272,$$

$$\ldots$$

$$r_{10} = 5.477225575051661.$$

$$(2.9.16)$$

The precision increases with the number of iterations, which depends also on how close is the "guessed" root from the real value. This is illustrated in Fig. 2.2 where we have chosen $r_0 = 1, 5, 15$. It is worth noting that if an approximation with six decimal places is considered acceptable, then the "convergence" to the desired value is ensured in a few iterations (4 or 5). The iteration converges to the asymptotic value[10] even if the initial conditions is very far from the actual root.

The method we have just outlined is a particular case of a more modern technique (tracing back to Newton) and called the method of the tangent, according to which the zero of a given function $f(x)$

---

[10]Namely the value obtained by a sufficiently large number of iterations.

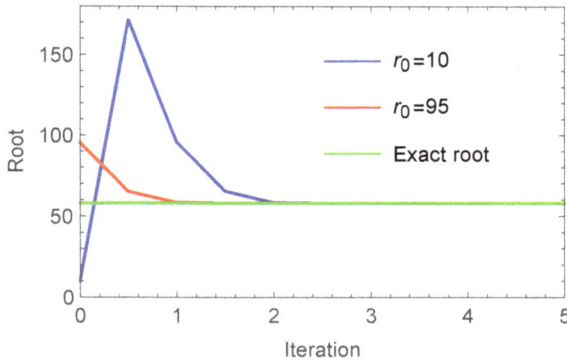

Figure 2.2: Iteration method for the square root of $R = 3333$ (Babylonian or Heron algorithm).

is obtained through the iteration

$$x_{k+1} = x_k + \frac{f(x_k)}{f'(x_k)}, \qquad\qquad f'(x) = \frac{d}{dx}f(x) \qquad (2.9.17)$$

By choosing $f(x) = x^2 - R$ (whose root is $x = \sqrt{R}$) we find $f'(x) = 2x$ and

$$x_{n+1} = x_n - \frac{x_n^2 - R}{2x_n} = x_n - \frac{x_n}{2} + \frac{R}{2x_n} = \frac{x_n}{2} + \frac{R}{2x_n} \qquad (2.9.18)$$

and therefore

$$x_{n+1} = \frac{1}{2}\left(x_n + \frac{R}{x_n}\right) \qquad (2.9.19)$$

which is the Heron method.

A further procedure aimed at justifying the Babylonian algorithm is offered by the following identity

$$2x^2 = x^2 + R. \qquad (2.9.20)$$

The solution of this second degree equation can be written as

$$x = \frac{1}{2}\left(x + \frac{R}{x}\right). \qquad (2.9.21)$$

The use of the approximation by successive iterations yields the square root algorithm.

An analogous procedure will be exploited in the forthcoming section to derive an iterative procedure to find the cubic roots.

Let us now consider the second degree equation

$$x^2 - x = N \qquad (2.9.22)$$

whose positive root can be written as

$$x_+ = \frac{1 + \sqrt{1 + 4N}}{2}. \qquad (2.9.23)$$

It is furthermore evident that if the original equation is written as

$$x = 1 + \frac{N}{x}, \qquad (2.9.24)$$

we can extend the Babylonian algorithm to the search of the relevant solution, using the following iterated step identity

$$x_{n+1} = 1 + \frac{N}{x_n}. \qquad (2.9.25)$$

In Fig. 2.3 we have reported the solution for $N = 7$, obtained after a certain number of iterative steps. The choice of $x_0 = 5$ yields the asymptotic value $3.1926\ldots$ (the positive solution of the second degree equation) after a limited number of steps.

Before closing this section we like to mention the following alternative to the Heron formula, according to which the square root of R can be computed as

$$\sqrt{R} = 2x^* - 1 \qquad (2.9.26)$$

where x^* is the asymptotic value of the recurrence

$$x_{n+1} = x_n + \frac{R - 1}{4x_n}, \qquad (2.9.27)$$

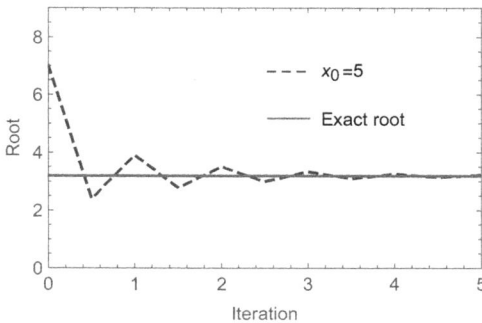

Figure 2.3: Iteration method for the second degree equation with $N = 7$.

whose reliability can be checked by comparison with the already discussed iteration (2.9.19).

A "practical" outcome of the previous discussion comes from the second iteration of the Heron algorithm, namely (see (eqs. (2.9.15)-(2.9.16))

$$\sqrt{R} \simeq r_2 = \frac{R^2 + 24r_0^2(R-1) + 16r_0^4 - 2R + 1}{8r_0(4r_0^2 + R - 1)} \qquad (2.9.28)$$

We can choose r_0 as the largest number (not necessarily an integer) whose square is closest to R (namely such that $r_0^2 < R$). We consider therefore that the accuracy of the approximation depends on the choice of r_0. Choosing for example $r_0 = 5$ and $R = 35$ the previous formula yields an approximation correct up to the fourth decimal place, while keeping $r_0 = 8$ and $R = 65$ the approximation is certainly much better and indeed we find an approximation valid at the ninth decimal digit. In the next section we will extend the method to study the derivation of higher order roots $(3, 4, 5, \dots)$ which were mastered by the ancient cultures.

2.10   Cubic Roots

In the previous section we have presented methods useful to derive the square roots using either hand or computer calculations. We apply here the same point of view to the cubic roots. We expose a fairly simple and effective procedure (known to ancient Babylonian and Indian Mathematicians) to evaluate the cubic roots of numbers up to six digits. The method we are going to introduce is a tricky procedure of interpolation and holds to derive roots of perfect cubes. The first step is the construction of the following table containing the cubes up to 9.

n	n^3
0	0
1	1
2	8
3	27
4	64
5	125
6	216
7	343
8	512
9	729

(2.10.1)

As preliminary example we compute $\sqrt[3]{5832}$. We group the radicand digits by separating (as shown below) the last three from the first. We write, under the first digit, the number whose cube is less than 5 (namely 1). We note that the last group of digits ends by 2 and check that the only cube ending by 2 is 512 corresponding to 8. It is put under 832, as shown below. It is finally checked that the cubic root we are interested in is simply 18:

$$
\sqrt[3]{5832} \rightarrow
\begin{matrix}
5 & 832 & - \\
1 & 8 & - \\
- & - & 18
\end{matrix}
$$

(2.10.2)

The following exercises are useful to familiarize with the protocol.

Exercise 4. *1.*

$$\sqrt[3]{19683} \rightarrow \begin{array}{ccc} 19 & 683 & - \\ 2 & 7 & - \\ - & - & 27 \end{array} \ . \tag{2.10.3}$$

2.

$$\sqrt[3]{117649} \rightarrow \begin{array}{ccc} 117 & 649 & - \\ 4 & 9 & - \\ - & - & 49 \end{array} \ . \tag{2.10.4}$$

3.

$$\sqrt[3]{8000} \rightarrow \begin{array}{ccc} 8 & 000 & - \\ 2 & 0 & - \\ - & - & 20 \end{array} \ . \tag{2.10.5}$$

We have underlined that the method does not work for non perfect roots, we emphasize the caveat here by noting that if applied to $\sqrt[3]{729729}$ we find 99 which is wrong! We invite the reader to find an argument justifying the procedure we have described.

Reliable approximate methods are therefore in order, the use of a generalization of the Babylonian algorithm can be applied by noting that

$$3x^3 = 2x^3 + R \rightarrow x = \frac{1}{3}\left(2x + \frac{R}{x^2}\right) \rightarrow x_{n+1} = \frac{1}{3}\left(2x_n + \frac{R}{x_n^2}\right), \tag{2.10.6}$$

which can be exploited using the same suggestions adopted for the square root (namely x_0 as close as possible to the cubic root of R to reduce the number of computational steps). In Fig. 2.4 we have reported an example of derivation of the cubic root of 30 by the use of the modified Heron method with two different initial values. The comments to be added are the same holding for the square root, with a modest caveat in addition: a uniform convergence to the solution is obtained by choosing an initial point larger than its exact value.

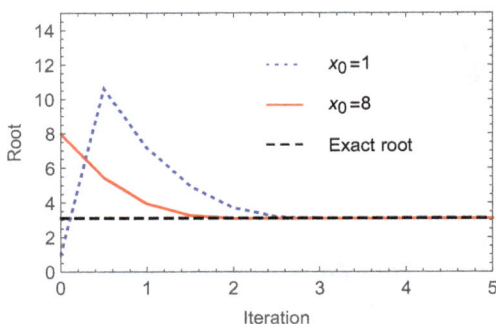

Figure 2.4: Iteration method for the cubic root of $R = 30$ (modified Heron method).

2.11 The Art of Computing and the Abacus Masters

The history of computation unfolds over five (and maybe more) millennia, its evolution is parallel to that of civilization itself. It is therefore not surprising that the algebraic problems were first put where humans began to develop social forms, which made use of advanced technologies and of complex commercial exchanges.

The following Fig. 2.5 shows a copy of a Sumerian tablet, dating back to the third millennium BC and that refers to a grain distribution problem.

Probably it is an accounting example to be used as a reference scheme. The data of the problem are the following: if you have a granary containing 1.152.000 $SILA$[11]) of wheat and you forage the population with 7 $SILA$ a person, how many will receive the same grain ration?

---

[11]Sumerian volumetric measure corresponding to about 8.3 liters.

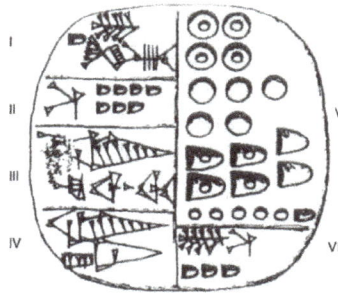

Figure 2.5: Sumerian tablet, 2650 B.C.

In the next Fig. 2.6 we refer to a Babylonian tablet dating from 1990 to 1600 BC in which we report a geometric problem, which is essentially the Pythagorean Theorem, asking for finding the diagonal of a square of side 30.

Figure 2.6: Babylonian tablet, 1990–1600 B.C.

The problem involves the square root of 2, a number that does not seem to have produced any scandal for the Babylonians, certainly more pragmatic than the Greeks towards mathematics, as we will see later in this book. The method of calculating the roots by the mathematicians of Babylon appears really worthy of note and we quoted "en passant" about the method of Heron. The calculation of $\sqrt{2}$ shown in the tablet in Fig. 2.6 is accurate to the sixth decimal place. The Sumerian used the sexagesimal basis and, translated from cuneiform characters, the number reads

$$\sqrt{2} \simeq 1; 24, 51, 10, \qquad\qquad 30\sqrt{2} \simeq 42; 25, 35. \qquad (2.11.1)$$

Let us now go back to the methods, developed at the beginning of mathematical thought, for the solution of algebraic equations. A papyrus tracing back to 2000 BC (named Rhind, after the art merchant who bought it) provides a kind of textbook manual containing among the other things a "general" method for the solutions of linear algebraic equation. One of the problems is *Evaluate the quantity which added to its seventh part yields* 19. Written in modern terms the problem corresponds to the search of the solution of the equation

$$x + \frac{x}{7} = 19 \qquad\qquad (2.11.2)$$

with the evident solution $x = \dfrac{7 \cdot 19}{8}$.

The Egyptians had been among the first to face the problem of fractions and related operations, therefore they had developed some arithmetical calculation skills and invented what was later called the rule of false position (the Latin term coined by Leonardo Fibonacci is *Regula Falsi*). The derivation of the exact solution in modern terms, even though at the level of any elementary school pupil, requires the ability to play with algebraic symbols. If you do not have this technique, you must sharpen the ingenuity and the idea behind the "Regula Falsi" is that of assigning a certain arbitrary (and therefore false) value to the "unknown" (*aha* for the Egyptians).

We choose, for this specific example, $x = 7$, in reality behind this assumption there is something more subtle because "7" does not imply 7 but seven units of a certain quantity unknown to us. Still in modern terms, indicating with u such reference unit we will have with $x = 7u$

$$7u + \frac{7u}{7} = 19 \Rightarrow u = \frac{19}{8} \Rightarrow x = 7 \cdot \frac{19}{8}. \qquad (2.11.3)$$

The strategy adopted by the Egyptians was to bring everything back to easily calculable units avoiding the search for least common multiple, that they evidently did not know. The result $x = 7 \cdot \frac{19}{8}$ is a decipherable expression for us but not for the Egyptians who needed something easily manageable in terms of integers or fractions, which can be expressed in multiples of 2. Therefore the procedure, to be effective, needs the further computational remark

$$7 = 4 + 2 + 1, \qquad 19 = 16 + 2 + 1,$$

$$\frac{19}{8} = \frac{16}{8} + \frac{2}{8} + \frac{1}{8} = 2 + \frac{1}{4} + \frac{1}{8},$$

$$7 \cdot \frac{19}{8} = 4 \left(2 + \frac{1}{4} + \frac{1}{8} \right) + 2 \left(2 + \frac{1}{4} + \frac{1}{8} \right) + \left(2 + \frac{1}{4} + \frac{1}{8} \right)$$

$$= 8 + 1 + \frac{1}{2} + 4 + \frac{1}{2} + \frac{1}{4} + 2 + \frac{1}{4} + \frac{1}{8} = 16 + \frac{1}{2} + \frac{1}{8}.$$

$$(2.11.4)$$

We have insisted on this example to highlight how a problem, nowadays within the capabilities of any junior school child, required a few millennia and a considerable intellectual effort to be solved.

Once, the rule of the "Holy Trinity" was taught at elementary schools, to solve problems of the type just discussed. The notions implicit in this procedure were the proportions and the reduction method to the unit, conceptually equivalent to the Regula Falsi. The rule was in use in Italy at times when mathematical formulas and conventions, though existing, were circulating within close circles of the initiates. Schools training people in computing art did not exist and the instrument of transmission was the oral tradition, with all the associated ambiguity. With the intensification of exchanges and the emergence of extensive commercial practices, the necessity for a reference became cogent. For this purpose Giovanni Sfortunati produced in 1534 the "Nuovo Lume (The New Light) - book of Arithmetic" whose subtitle is truly illuminating (we omit the original version, written in ancient Italian) "*The Title is New Light, because many false propositions derived by former Authors are corrected and condemned*".

We insisted on this last remark to highlight that, towards the end of the Middle Ages, a bourgeois class emerged in southern Europe (mainly in the Mediterranean Area) which would have changed Western culture. They felt the need to introduce the so-called abacus workshops intended to train a new ruling class composed of merchants, engineers, architects ... to support and direct the new course of history. Practical problems like the composed interests for loan and commercial exchanges where at the time of paramount importance, more or like the sophisticated codes nowadays in use to predict financial market evolution. The extraction of the nth roots has been faced by the abacus masters of those years (around the 13th century) they developed approximate methods not unlike those illustrated so far.

Exercise 5. *We invite the reader to meditate on the following procedure for **root extraction** due to **Gerolamo Cardano**, which in modern mathematical language reads*

$$\sqrt[n]{N} \simeq a + b, \qquad\qquad b = \frac{N - a^n}{\sum_{k=1}^{n-1} \frac{n!}{k!(n-k)!} a^k} \qquad (2.11.5)$$

where a is the number better approximating N ($a^n < N$). In the following Fig. 2.7 we have reported a comparison between the $\sqrt[n]{100}$ ($n = 2, \dots, 10$) obtained with the Cardano formula and a nine digit result from a pocket calculator.

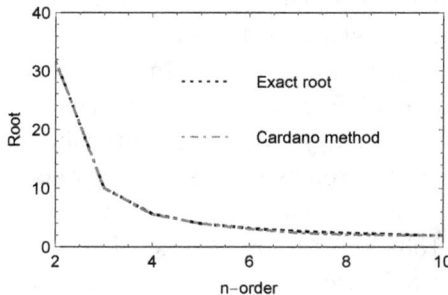

Figure 2.7: Iteration method for the n-order root of $N = 1000$, (Cardano formula).

This example completes the second Chapter of this book, in which we have discussed notions of elementary arithmetic. In the forthcoming chapters we will touch again some of these problems, within the context of a more elaborated perspective.

Chapter 3

Continued Fractions, Nested Radicals, Fibonacci, Pell ... and Transcendental Numbers

3.1 A Preliminary Introduction to Continued Fractions and to Nested Radicals

Do you know what is said to be incestuous in Mathematics? If not, you can look at the paper by J.D. Barrow "Chaos in Numberland"[1] where the author proposed the solution of the second degree algebraic equation $x^2 - x - N = 0$ as

$$x = 1 + \frac{N}{x} \qquad (3.1.1)$$

which appears an attempt of generating an unknown by the unknown itself. The mix of the same genus may have suggested this rather crude, but effective, image.

---

[1]It is available on http://plus.maths.org/content/chaos-cumberland-secret-life-continued-fractions.

In the previous chapter we have introduced relations of such a type to infer an iterative procedure to evaluate square or cubic roots, without raising any moral prejudice. If we look at the above equation as a recurrence, namely as

$$x_{n+1} = 1 + \frac{N}{x_n},$$ (3.1.2)

we obtain the following solutions at successive iterations

$$1 + \frac{N}{1 + \frac{N}{x_1}} \rightarrow 1 + \frac{N}{1 + \frac{N}{1 + \frac{N}{x_2}}} \rightarrow 1 + \frac{N}{1 + \frac{N}{1 + \frac{N}{x_3}}} \rightarrow \dots$$ (3.1.3)

This is what is called an *Infinite continued fraction* (*ICF*) and can be exploited to provide, for example, expansions of irrational numbers.

An example of *Limited continued fractions* (*LCF*) is provided by

$$2 + \frac{1}{3 + \frac{1}{4 + \frac{1}{2}}}.$$ (3.1.4)

They leads to rational numbers (in this specific case $\frac{67}{29}$).

The interest for this type of CF is limited. It should however be noted that any rational number is reducible to a LCF, as shown by the following examples [2]

$$\frac{6}{19} = \frac{1}{\frac{19}{6}} = \frac{1}{3 + \frac{1}{6}},$$

$$\frac{57}{179} = \frac{1}{\frac{179}{57}} = \frac{1}{3 + \frac{8}{57}} = \frac{1}{3 + \frac{1}{\frac{57}{8}}} = \frac{1}{3 + \frac{1}{7 + \frac{1}{8}}}.$$ (3.1.5)

The trick we have exploited is that of splitting the denominator in the sum of an integer and a fraction whose nominator is 1. We will

---

[2]We will see in the following that this definition is not fully correct because it is not a "canonical", we exploit this example because it yields an idea of the forms we utilize in the following.

also see later that an ICF truncated into a LCF represents the rational approximation of an irrational number.

We introduce the "nested radicals" by proposing the following problem: *"What is the number represented by the following repeated radicals?"*

$$\sqrt[3]{10 - \sqrt[3]{10 - \sqrt[3]{10 - \ldots}}} \qquad (3.1.6)$$

The strategy we may follow to provide an answer, is illustrated below.

a) Set

$$x = \sqrt[3]{10 - \sqrt[3]{10 - \sqrt[3]{10 - \ldots}}} \; ; \qquad (3.1.7)$$

b) Keep the cube of both sides and get

$$x^3 = 10 - \sqrt[3]{10 - \sqrt[3]{10 - \ldots}} = 10 - x \; ; \qquad (3.1.8)$$

c) Note that one of the solutions of the cubic equation $x^3 + x = 10$ is $x = 2$. Accordingly

$$\sqrt[3]{10 - \sqrt[3]{10 - \sqrt[3]{10 - \ldots}}} = 2 \; . \qquad (3.1.9)$$

Let us now consider the solution of eq. (3.1.1), written as $x = \sqrt{N + x}$, which can be exploited to write the recurrence

$$x_{n+1} = \sqrt{N + x_n} \; \rightarrow$$

$$\rightarrow \sqrt{N + x_1} \; \rightarrow \sqrt{N + \sqrt{N + x_2}} \; \rightarrow$$

$$\rightarrow \sqrt{N + \sqrt{N + \sqrt{N + x_3}}} \; \rightarrow \sqrt{N + \sqrt{N + \sqrt{N + \sqrt{N + x_4}}}} \; \rightarrow \ldots$$

$$(3.1.10)$$

representing the solution of a second degree algebraic equation in terms of *Nested unlimited radicals* (NUR)[3].

---

[3] Their form suggests the definition of nested radicals, but also iterated radicals is a correct identification.

The comparison between NUR and ICF solutions is provided in Fig. 3.1. It might be argued that the procedure, we have described, holds for real solutions only, this is not the case and in Fig. 3.2 we consider the iteration for equations admitting complex roots.

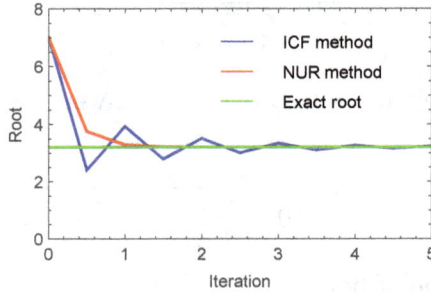

Figure 3.1: Iteration methods for the second degree equation with $N = 7$.

The methods we have just outlined allow the derivation of a number of funny identities (not useful anytime). We invite the reader to prove that the following expansions in terms of NUR and ICF hold

$$R = 2\sqrt{\frac{R^2 - 1}{4} + \sqrt{\frac{R^2 - 1}{4} + \sqrt{\frac{R^2 - 1}{4} + \ldots}} - 1},$$

$$R = 1 + 2\cfrac{\frac{R^2 - 1}{4}}{1 + \cfrac{\frac{R^2 - 1}{4}}{1 + \cfrac{R^2 - 1}{4}}{1 + \ldots}} . \tag{3.1.11}$$

The NUR expansion is unpractical to approximate square roots (instead of one square root one calculates many), the second (once truncated) requires a finite number of rational approximations. The ICF expansion is not convenient too, because very slowly converging, thus a large number of terms need to be evaluated before getting a reasonable approximation.

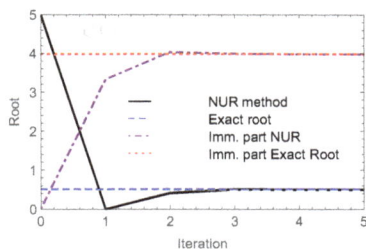

(a) Iteration methods NUR. (b) Iteration methods ICF.

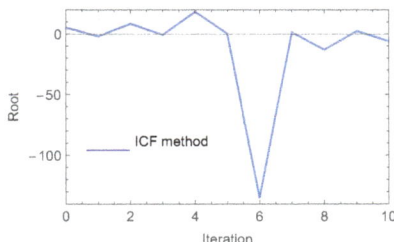

Figure 3.2: Iteration methods for the second degree equation with $N = 16$ and complex roots.

We conclude this introductory section by proposing a further exercise, namely by inviting the reader to prove the following statement

$$\frac{2}{\pi} = \prod_{n=0}^{\infty} \frac{x_n}{2}, \qquad x_{n+1} = \sqrt{1 + x_n}, \qquad x_0 = 0 \quad (3.1.12)$$

which eventually writes

$$\frac{2}{\pi} \simeq \frac{\sqrt{2}}{2} \frac{\sqrt{2 + \sqrt{2}}}{2} \frac{\sqrt{2 + \sqrt{2 + \sqrt{2}}}}{2} \cdots \quad (3.1.13)$$

known as **Vieté Formula**.

3.2 Irrational Numbers and Continued Fractions

We have so far denoted continued fractions by the use of a sloppy and cumbersome and misleading notation, in the following we adopt the alternative expression (with $a_i \in \mathbb{Z}^+$)

$$[a_0; a_1, a_2, \ldots, a_n] = a_0 + \cfrac{1}{a_1 + \cfrac{1}{a_2 + \cfrac{\cdots}{\cdots + \cfrac{1}{a_n + \frac{1}{\cdots}}}}} \quad (3.2.1)$$

Accordingly, the solution in terms of a ICF of the second degree algebraic equation

$$x^2 = ax + 1 \qquad (3.2.2)$$

can be written as

$$x_+ = \frac{a + \sqrt{a^2 + 4}}{2} = a + \cfrac{1}{a + \cfrac{1}{a + \cdots}} = [a; a, a, \ldots]. \qquad (3.2.3)$$

It is therefore evident that the ***golden number***

$$\phi = \frac{1 + \sqrt{5}}{2} \qquad (3.2.4)$$

can be written in terms of ICF as

$$\left.\frac{a + \sqrt{a^2 + 4}}{2}\right|_{a=1} = [1; 1, 1, \ldots]. \qquad (3.2.5)$$

It is now fairly natural to exploit the continued fraction formalism to express the irrational numbers in terms of ICF. For future convenience, we remind that a number is irrational if its expansion (in any base) consists of infinite digits without any periodicity.

The first example is the square root of 2 which is expanded in ICF by use of the following procedure

$$x^2 = 2 \; \rightarrow \; x^2 - 1 = 1 \; \rightarrow \; (x+1)(x-1) = 1 \; \rightarrow \; x = 1 + \frac{1}{1+x}$$

$$\rightarrow \; 1 + \cfrac{1}{2 + \cfrac{1}{2 + \cfrac{1}{2 + \cdots}}} .$$

$$(3.2.6)$$

It is therefore evident that

$$\sqrt{2} = [1; 2, 2, 2, \ldots] = [1; \bar{2}], \qquad (3.2.7)$$

where $\bar{2}$ denotes the periodicity of the ICF, namely that 2 is repeated an infinite number of times.

As already anticipated, we note that by truncating the expansion we can get an approximation (at various orders of the root itself). An approximation of $\sqrt{2}$ appearing in the vedic books is given (in modern terms) below,

$$\sqrt{2} \simeq 1 + \frac{1}{3} + \frac{1}{3 \cdot 4} + \cdots \qquad (3.2.8)$$

The mystery surrounding its provenience is unveiled by keeping the third order expansion of the ICF, which yields

$$\sqrt{2} \simeq [1; 2, 2] = 1 + \frac{1}{2 + \dfrac{1}{3}} = 1 + \frac{1}{2}\left(1 - \frac{1}{6}\right) = 1 + \frac{1}{3} + \frac{1}{3 \cdot 4}. \quad (3.2.9)$$

The inclusion of a further term in the expansion yields the further correcting term, the vedic approximation can indeed be written as $\sqrt{2} \simeq [1; 2, 2, 2]$.

We will further comment in the following the importance of the continued fractions to approximate different types of problems.

In the case of $\sqrt{3}$ we use the same procedure and find

$$x^2 = 3 \rightarrow x^2 - 1 = 2 \rightarrow (x+1)(x-1) = 2 \rightarrow x = 1 + \frac{2}{1+x}$$

$$\rightarrow 1 + \frac{2}{1 + 1 + \frac{2}{1+x}} \rightarrow 1 + \frac{2}{2 + \frac{2}{1+1+\frac{2}{1+x}}}$$

$$\rightarrow \cdots \rightarrow 1 + \frac{1}{1 + \frac{1}{2 + \frac{1}{1+\frac{1}{1+x}}}} \rightarrow \cdots \rightarrow [1; 1, 2, 1, 2, \ldots] \rightarrow [1; \overline{1, 2}].$$

$$(3.2.10)$$

Regarding $\sqrt{5}$ we get

$$\sqrt{5} = [2; \overline{4}]. \qquad (3.2.11)$$

The expansion in continued fractions is a matter of experience and hence of exercises. We propose therefore the following exercises to check the level of confidence in this matter.

Example 11. *We consider here the following irrational $\sqrt{n^2+1}$ (here and below $n \in \mathbb{Z}$) and look for a procedure to get an expansion independently of the specific values of n. To this aim we note that*

$$n + \cfrac{1}{n+\sqrt{n^2+1}} = n + \cfrac{n - \sqrt{n^2+1}}{\left(n + \sqrt{n^2+1}\right)\left(n - \sqrt{n^2+1}\right)} = \sqrt{n^2+1}.$$

$$(3.2.12)$$

According to the above trick, the 95% of the game is done. It follows indeed that

$$\sqrt{n^2+1} = n + \cfrac{1}{n+\sqrt{n^2+1}} = n + \cfrac{1}{2n+\cfrac{1}{n+\sqrt{n^2+1}}} = n + \cfrac{1}{2n+\cfrac{1}{2n+\cfrac{1}{n+\sqrt{n^2+1}}}}$$

$$= \ldots = [n; \overline{2n}].$$

$$(3.2.13)$$

The use of the same procedure allows to derive the further identity

$$\sqrt{n^2+2} = [n; \overline{n, 2n}]. \qquad (3.2.14)$$

Exercise 6.   *1.*

$$\sqrt{15} = [3; \overline{1, 6}], \qquad (3.2.15)$$

2.

$$\sqrt{31} = [5; \overline{1, 1, 3, 5, 3, 1, 1, 1, 10}], \qquad (3.2.16)$$

3. The expansion in ICF of the square root of 31 is quite laborious but it is a very good training effort to test the level of ability achieved in the handling of the formalism.

According to the previous discussion it is easily checked that any continued fraction of the type $[n; \overline{m, p}]$ can be expressed as the solution of a second degree equation which can be derived from

$$x = n + \cfrac{1}{m + \cfrac{1}{p-n+x}}. \qquad (3.2.17)$$

We invite the reader to solve the associated algebraic equation and speculate on the possibility of exploiting this result to get expansions like

$$\sqrt{8} = [2; \overline{1, 4}], \qquad\qquad \sqrt{18} = [4; \overline{4, 8}] \qquad (3.2.18)$$

and also to think about the following identity

$$\sqrt{8} = 2\sqrt{2} = 2[1;\overline{2}] = [2;\overline{1,4}] \qquad (3.2.19)$$

and find a suggestion specifying a criterion to multiply an integer by a *ICF* and get another *ICF*.

We like to observe that an irrational number has an expansion in terms of a periodic ICF, while its expansion in terms of a decimal basis (or p as well) is unlimited and a-periodic.

Talking about transcendental numbers, we dwell on this point later in this chapter.

A Practical Rule: The square root of any number R expressible as

$$R = p^2 + 1 \qquad (3.2.20)$$

can be written as

$$\sqrt{R} = [p;\overline{2p}]. \qquad (3.2.21)$$

The relevant proof goes as it follows

$$x^2 = R = p^2 + 1 \rightarrow (x+p)(x-p) = 1 \rightarrow x = p + \frac{1}{p+x}$$

$$\rightarrow p + \frac{1}{p+p+\frac{1}{p+x}} \rightarrow \cdots \rightarrow p + \frac{1}{2p + \frac{1}{2p + \frac{1}{2p + \frac{1}{p+x}}}}$$

$$\rightarrow \cdots \rightarrow [p; 2p, 2p, \ldots] \rightarrow [p; \overline{2p}].$$

$$(3.2.22)$$

A periodic *ICF* can be truncated at a given order, namely

$$[a_0; a_1, a_2, \ldots, a_n, \ldots] \simeq \begin{array}{c} [a_0] \\ [a_0; a_1] \\ [a_0; a_1, a_2] \\ \cdots \end{array} \qquad (3.2.23)$$

which represent the $0th$, $1st$, $2nd$, ... level of approximation respectively.

We have already noted that the truncation of a periodic ICF, at a given order, returns a rational approximation of the corresponding irrational number. We provide an example clarifying the level of the approximation which can be achieved:

$$\sqrt{1+x^2} = [x; \overline{2x}] = \begin{cases} x + \dfrac{2x}{4x^2+1} & \text{order 2} \\[4mm] x + \dfrac{4x^2+1}{8x^3+4x} & \text{order 3} \end{cases}. \qquad (3.2.24)$$

The approximation improves for increasing values of x and in Fig. 3.3 we have reported different levels of approximations, which hold for orders $2, 3, 4$. For numbers above 1 the error is less than 1% even for the second order case as it is shown in Fig. 3.4.

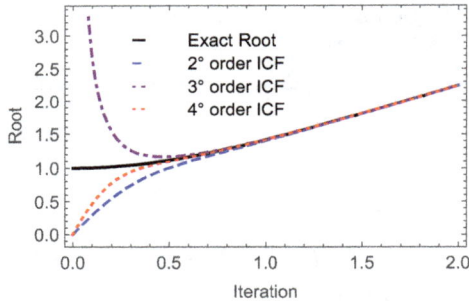

Figure 3.3: Comparison through different orders of ICF method for $\sqrt{1+x^2}$.

A nice and instructive exercise is the use of the previous formalism to show that the rational number $\frac{4756}{3363}$ approximates $\sqrt{2}$ at the sixth decimal digit (see below). We have loosely defined the *order of the truncation of an ICF* the *LCF* obtained by keeping an increasing

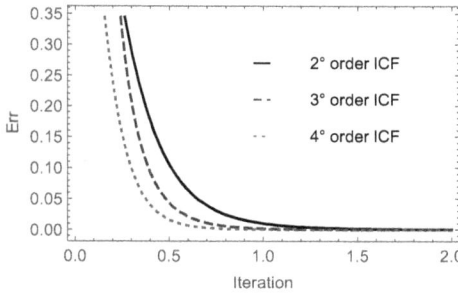

Figure 3.4: Relative error (%) for different orders of ICF method for $\sqrt{1+x^2}$.

number of terms a_i and progressively arranging them as

$$[a_0; a_1, a_2, ..., a_n, ...] \rightarrow$$

$$
\begin{aligned}
[a_0] &= a_0 \\
[a_0; a_1] &= \frac{a_0 a_1 + 1}{a_1} \\
[a_0; a_1, a_2] &= \frac{a_2(a_0 a_1 + 1) + a_0}{a_2 a_1 + 1} \\
[a_0; a_1, a_2, a_3] &= \frac{a_3[a_2(a_0 a_1 + 1) + a_0] + a_0 a_1 + 1}{a_3(a_2 a_1 + 1) + a_1} \\
\cdots &= \cdots \\
[a_0; a_1, a_2, ..., a_n, ...] &= \cdots
\end{aligned}
$$

$$(3.2.25)$$

In more educated terms they are called *convergents* and can always be written as $\frac{p_n}{q_n}$. It is easily checked that they satisfy the recurrences

$$
\begin{aligned}
p_n &= a_n p_{n-1} + p_{n-2}, & p_{-2} &= 0, & p_{-1} &= 1, & p_0 &= 1, \\
q_n &= a_n q_{n-1} + q_{n-2}, & q_{-2} &= 1, & q_{-1} &= 0, & q_0 &= 1
\end{aligned}
$$

$$(3.2.26)$$

and therefore we find

$$[a_0; a_1, a_2, \ldots, a_n] = \frac{p_n}{q_n} = \frac{a_n p_{n-1} + p_{n-2}}{a_n q_{n-1} + q_{n-2}}. \qquad (3.2.27)$$

The convergents of a periodic ICF are easily specified, the term a_n is indeed easily defined. In the case of $\sqrt{2}$ ($a_n = 2$) the relevant rational approximation for $n = 8$ is provided by the fraction $\frac{577}{408}$ which is good up to the fifth decimal digit. In the case of $\sqrt{5}$ ($a_n = 4$) the fourth order approximation is $\frac{161}{172}$ with an error of the order[4] 0.001.

A historical example of application of continued fraction is the relevant use in the reform of calendar. Pope Gregorio XIII proposed a commission for the reform of the Julian calendar. The reason was due to the fact that the calendar proposed by Julius Caesar (45 *BC*) was based on an approximation of the period of revolution of the Earth around the Sun, which led to an error which cumulated in the centuries, in such a way that in 1582 (the year of the reform) the error was of about 10 days in advance. Before proceeding, we just remind that if we consider a number, known in its decimal expansion, say 3.14 we can write its (finite) *CF* as

$$3,14 = 3 + \frac{14}{100} = 3 + \frac{1}{\frac{50}{7}} = 3 + \frac{1}{7 + \frac{1}{7}} = [3; 7, 7]. \qquad (3.2.28)$$

Let us consider the period of terrestrial rotation around the Sun measured in days and fraction of days

$$T \simeq 365, 2422189. \qquad (3.2.29)$$

Making a boring computation we end up with

$$T \simeq [365; 4, 7, 1, 3, 40, 2, 3, 5]. \qquad (3.2.30)$$

The first order approximation reads

$$T \simeq 365 + \frac{1}{4} \qquad (3.2.31)$$

which suggests that to better approximate the period with respect the Egyptian assumption of 365 days, it will be sufficient to add one day every four years, according to the rule:

---

[4]There are many noticeable essay in Math literature on *CF*, some of them are hardly useful for non mathematicians. Many appreciable efforts appear on the web too. A contribution we have very much appreciated can be found at http://www.maths.surrey.ac.uk/hosted-sites/R.Knott/Fibonacci/cfINTRO.html#section9.

Each year that is divisible by 4 consists of 366 days and each other year consists of 365 days.

This was the basis of the **Julian calendar**. This approximation was not accurate even for that historical period, it was clear for the astronomer at that time that this approximation leads to one day in advance in 136 years, consistent with an approximation of 0.7%.

The second approximation yields

$$T \simeq 365 + \frac{8}{33} \tag{3.2.32}$$

which suggests that any 33 years, 8 days should be added.

During the 11th century, the Persian mathematician Omar Khayyam (1048−1131) suggested a 33-year cycle where the years 4, 8, 12, 16, 20, 24, 28 and 33 should be leap years. Thus the mean-duration of a year according to his suggestion would be exactly the value of the third approximation, of the CF in eq. (3.2.30).

The further approximations are

$$T \simeq 365 + \frac{31}{128}, \qquad T \simeq 365 + \frac{1248}{5153}, \qquad T \simeq 365 + \frac{3775}{15587}.$$
$$\tag{3.2.33}$$

The last two are not practical and to get a compromise the **Gregorian calendar** is based on the following approximation

$$T \simeq 365 + \frac{97}{400}, \tag{3.2.34}$$

namely 97 days to be added any 400 years. In order to fulfill this condition, the leap years convention is:

1) The year is evenly divisible by 4;
2) If the year can be evenly divided by 100, it is *not* a leap year, unless;
3) The year is also evenly divisible by 400: then it is a leap year.

The Gregorian reform (actually due to the mathematician **Luigi Lillo**) allowed the realignment of the calendar with astronomical events like equinoxes and solstices.

The rules for calculating the Christian Easter dates can be worked out using analogous procedures. A general formula was not an easy task and had to be obtained by successive approximations. An attempt in this direction was made by the great German mathematician **Karl Friedrich Gauss**, successively it was refined in a more general context by Spencer Jones[5] in 1922. We provide (without any proof) the **Algorithm for the Easter Date** (valid for the Gregorian Calendar) relevant to the year X. The protocol is given below:

$$a = X - 19\lfloor\frac{X}{19}\rfloor, \qquad b = \lfloor\frac{X}{100}\rfloor, \qquad c = X - 100b, \qquad d = \lfloor\frac{b}{4}\rfloor,$$

$$e = b - 4d, \qquad f = \lfloor\frac{b+8}{25}\rfloor, \qquad g = \lfloor\frac{b-f+1}{3}\rfloor,$$

$$h = 19a + b - d - g + 15 - 30\lfloor\frac{19a + b - d - g + 15}{30}\rfloor, \qquad i = \lfloor\frac{c}{4}\rfloor,$$

$$k = c - 4i, \qquad l = 32 + 2e + 2i - h - k - 7\lfloor\frac{32 + 2e + 2i - h - k}{7}\rfloor,$$

$$m = \lfloor\frac{a + 11h + 22l}{451}\rfloor, \qquad n = \lfloor\frac{h + l - 7m + 114}{31}\rfloor,$$

$$p = h + l - 7m + 114 - 31n.$$

$$(3.2.35)$$

You may trust or not, but the result of the computation is

$$\binom{month}{day} = \binom{n}{p+1}. \qquad (3.2.36)$$

Specifying month and day at which Easter occurs in the year X, e.g

$$1945 \rightarrow \binom{4}{1}, \qquad 2020 \rightarrow \binom{4}{12}, \qquad 2035 \rightarrow \binom{3}{25}, \qquad (3.2.37)$$

which can be checked with a corresponding calendar.

---

[5]For further comments see J. Meeus "Astronomical Formulae for Calculators", ISBN: 0943396220,9780943396224.

3.3 Recurrence Equations, Pell and Fibonacci Numbers

A *Recurrence equation* (RE) is any identity of the type

$$k_{n+1} = f(k_n), \qquad n = 0, 1, 2, \ldots \qquad (3.3.1)$$

in which the unknown is k_n which is a sequence of numbers specified by the integer n.

The equations mentioned in the previous section for the definition of the convergents are therefore RE. The relevant solutions are in general not easily found, except particular, but important, cases. We provide a glimpse on the associated method of solutions by considering the following example in which the coefficients of the equation are not dependent on the index n, namely

$$s_n = a s_{n-1} + s_{n-2}, \qquad s_0 = \alpha, \qquad s_1 = \beta. \qquad (3.3.2)$$

The conditions on the first two terms of the unknown $0, 1$ (or any other two terms) are necessary to get a not undetermined form. This is similar to what occurs in the search of solutions of second order ordinary differential equations, with constant coefficients. We use the **Binet method** and look for a general solution by setting

$$s_n = s^n \qquad (3.3.3)$$

plugging it in the original equation. We reduce the problem to the search of the roots of a second degree algebraic equation

$$s^2 - as - 1 = 0. \qquad (3.3.4)$$

The RE is accordingly solved by embedding the partial solutions involving the $+, -$ roots in the following linear combination

$$s_n = l s_+^n + m s_-^n \qquad (3.3.5)$$

whose arbitrary constants can be fixed through the "initial conditions" which yield

$$l + m = \alpha, \qquad l s_+^1 + m s_-^1 = \beta \qquad (3.3.6)$$

and eventually finding

$$s_n = \frac{1}{s_+ - s_-} \left[\alpha \left(s_- s_+^n - s_+ s_-^n \right) - \beta \left(s_+^n - s_-^n \right) \right]. \qquad (3.3.7)$$

We specialize our result to the case in which $a = 2$, $s_0 = 0$, $s_1 = 1$ and obtain the following result

$$p_n = \frac{\left(1 + \sqrt{2}\right)^n - \left(1 - \sqrt{2}\right)^n}{2\sqrt{2}}. \qquad (3.3.8)$$

The numbers generated by the recurrence are called **Pell numbers**[6] and are $0, 1, 2, 5, 12, 29, 70, 169, 408 \ldots$ It is worth noting that the ratio $R_n = \frac{p_{n+1}}{p_n} - 1$ represents a rational approximation of $\sqrt{2}$, which gets better with increasing n (e.g. $\frac{p_9}{p_8} - 1 = \frac{577}{408} = 1.414215686$, $\sqrt{2} \simeq 1.414213562373095$).

Assuming furthermore $a = 2p$, the associated recurrence is

$$s_n = 2 \, p \, s_{n-1} + s_{n-2}, \qquad\qquad s_0 = 0, \qquad\qquad s_1 = 1. \qquad (3.3.9)$$

The relevant solution reads

$$p_n = \frac{\left(1 + \sqrt{p^2 + 1}\right)^n - \left(1 - \sqrt{p^2 + 1}\right)^n}{2\sqrt{p^2 + 1}}. \qquad (3.3.10)$$

As shown in Fig. 3.5 the ratio R_n can be exploited to get successive approximations of $\sqrt{p^2 + 1}$.

The so-called **Fibonacci numbers**[7] (A000045) are a particular case of the Pell families and are written in terms of the mentioned

---

[6]In Online Electronic Integer Sequences "*OEIS*", the Pell numbers correspond to the code A002203. *OEIS* is a kind of Paradise for the lovers of numbers and series. Almost all known series and approximations and relevant references appear in there. We warmly suggest the reader (if belonging to the number-lover species) to give a look at this tool.

[7]See the Online Encyclopedia of Integer Sequences on http://oeis.org/A000045.

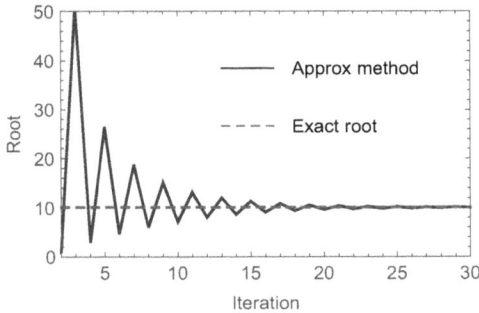

Figure 3.5: Rational approximation method for $\sqrt{p^2 + 1}$ with $p = 10$.

golden number (3.2.4) (or golden section) $\phi = \dfrac{1 + \sqrt{5}}{2}$, namely

$$F_n = \frac{1}{\sqrt{5}} \left[\left(\frac{1 + \sqrt{5}}{2} \right)^n - \left(\frac{1 - \sqrt{5}}{2} \right)^n \right] \qquad (3.3.11)$$

and are derived through the sequence

$$F_{n+2} = F_{n+1} + F_n, \qquad\qquad F_0 = 0, \qquad\qquad F_1 = 1. \qquad (3.3.12)$$

The fame of these numbers, which are cloaked in an esoteric aspect (not for the present Authors) owe their fame to the fact that they are expressed through the golden number, which has raised, since the antiquity, an interest mostly of platonic and mystic nature. On the other side, it is not clear why the so-called *Silver section*

$$\sigma = 1 + \sqrt{2} \;\Rightarrow\; p_n = \frac{\sigma^n - (1 - \sigma)^n}{2\sqrt{2}} \qquad (3.3.13)$$

and the associated Pell-like numbers are not endorsed with the same mystical flavour. In the following Figures 3.6 and 3.7 we report the geometrical construction of the **Silver and Golden sections**.

Fibonacci numbers enjoy an extraordinary credit, they have been called *Fabolous* and very often are quoted out of turn. They are (apparently) recurring in a number of problems (math, physics, biology,

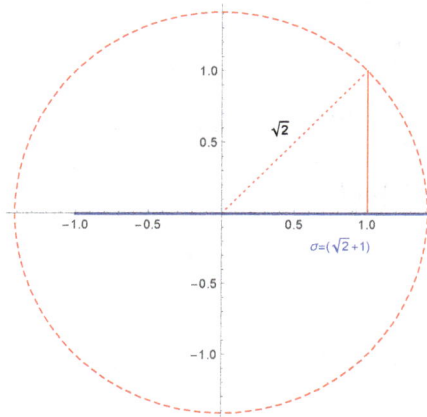

Figure 3.6: Silver section.

finance ...), but they are only a particular case of Pell numbers (which are much less appreciated) and both (Pell and Fibonacci) are expressed in terms of the quadratic irrationals, which will be discussed in the forthcoming section.

3.4   Quadratic Irrationals

The following couple of numbers

$$\alpha_+ = a + \sqrt{D}, \qquad\qquad \alpha_- = a - \sqrt{D} \qquad (3.4.1)$$

with $a, D \in \mathbb{Z}$, are *reciprocally conjugated quadratic irrational* ($RCQI$). They are said reduced if the conjugated α_- lies between the interval -1 and 0 (extremes excluded). The number $\alpha_+ = 1 + \sqrt{2}$ solution of the equation $x^2 - 2x - 1 = 0$ is a reduced $RCQI$. Numbers of the type

$$\alpha_+ = \frac{a + \sqrt{D}}{2}, \qquad\qquad \alpha_- = \frac{a - \sqrt{D}}{2} \qquad (3.4.2)$$

are QI if a is odd and[8] $D = 4p + 1$.

---

[8]Namely if D is congruent 1 mod 4.

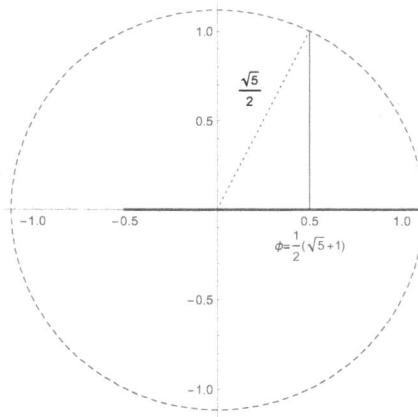

Figure 3.7: Golden section.

The number $\rho = \dfrac{3+\sqrt{5}}{2}$, called ***Bronze section*** is an $RCQI$ expressible in terms of the golden number as[9]

$$\rho = 1 + \phi. \tag{3.4.3}$$

All QI admits an expansion in terms of ICF[10], the root of numbers like $4p+1$ can be therefore expanded in ICF, thus admitting a rational approximation in terms of the associated LCF.

3.5 A Neo Platonic Intermezzo

Fibonacci numbers are a sequence of numbers with properties no more interesting than others of the same type. Many of the associated fascinations are mere misunderstandings and trace back to the golden section, a number called by the ancient mathematicians (but not by Greeks) *divine proportion*. We will further comment on these recurring shortcomings later in this book, we like here to remind that the golden section emerges from a problem proposed by

---

[9]$\rho = 1 + \phi$ is a quadratic irrational, even though not reduced.
[10]This statement is known as ***Galois Theorem***.

Euclid[11], namely (see also the following Fig. 3.8):

$$Given\ a\ segment\ \overline{AE}\ find\ a\ point\ B\ such\ that\ \frac{\overline{AE}}{\overline{AB}} = \frac{\overline{AB}}{\overline{BE}}.$$

We have no difficulty to provide the relevant solution as indicated below

$$\overline{MC} = \frac{\sqrt{5}}{2}, \qquad \overline{MB} = \frac{1}{2}, \qquad \overline{BE} = \frac{\sqrt{5}-1}{2}, \qquad (3.5.1)$$

and

$$\overline{AE} = \overline{AB} + \overline{BE} = \frac{\sqrt{5}+1}{2}, \qquad (3.5.2)$$

the Euclid problem is therefore solved. Fig. 3.8 reports the geometrical meaning of the golden ratio along with the construction of the golden rectangle.

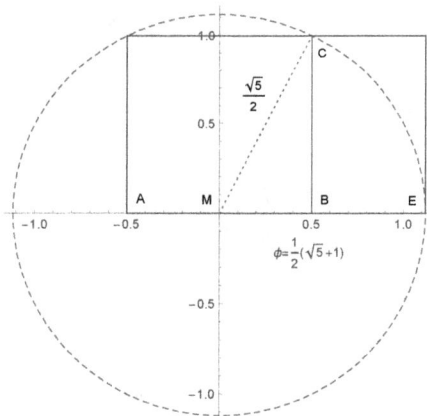

Figure 3.8: Golden rectangle construction.

The neo-platonic suggestions are hard to die and nestle in the mind of some scientist as perverse fantasies. It happens, for example, that someone has the dream or better the tendency of reducing Physics (and other aspects of science) to a kind of alchemic game.

---

[11]Elements, Liber VI 30th proposition.

There is the pretention of reducing the universal constants to combinations of numbers with special properties (prime, transcendental, irrational ...) and ϕ is one of them. Within the framework of certain theories[12], hardly reconciled with those accepted by the scientific community[13], it has been suggested that the mass of quarks are expressible in terms of the golden number as

$$m_u = \phi^3 + 1 = 5.2360679775 \ MeV,$$
$$m_d = 2\phi^3 = 8.472135954 \ MeV, \qquad (3.5.3)$$
$$m_s = 10\phi^6 = 179.442719 \ MeV,$$

Extreme cautions are necessary when dealing with this type of identities. At the moment they should be considered as numerical coincidences, albeit the quoted paper by Naschie has undoubtful reasons of interest. Further comments concerning these coincidences are given in the concluding chapter of the book.

Remaining within the same neo-platonic context, we quote the Titius–Bode "law", which links the distances of the planet from the sun, through the formula

$$d_n = \frac{1}{10}(2^{n-2}\,3 + 4), \qquad n = 2, 3, \ldots \qquad (3.5.4)$$

with d_n being a quantity expressed in terms of the Sun-Earth distance.

The comparison between formula predictions and the actual distances are given in Table 3.1 which contains a comparison with an alternative to the **Titius–Bode law** and provided by the **McLaughlin formula**

$$d_n = \frac{1}{10}(q_{n+3} + 3), \qquad n = 1, 2, \ldots \qquad (3.5.5)$$

---

[12]See e.g. M.S. El Naschie, "A review of E-infinite theories and mass spectrum of high energy particle physics", Chaos Solitons and Fractals n. 19, 2004, pp. 209–236. We are not expressing any negative opinion on the value of this paper, which has undoubtful merits and provides a radically different point of view with respect to the currently accepted doctrines.

[13]Quarks are the elementary constituents of matter, they are designed as u, d, s, c, b, t.

where q_n are **Tetranacci numbers** specified by the recurrence

$$q_n = q_{n-4} + q_{n-3} + q_{n-2} + q_{n-1}, \qquad q_0 = 0, \qquad q_{r \leq 4} = 1.$$
$$(3.5.6)$$

The reasons underlying the coincidence between formulae and measured values are not clear, we will just spend a few comments in the last chapter.

This example was also aimed at stressing that along with Pell numbers other families of numbers associated to non quadratic irrationals do exist.

We have quoted Tetranacci which are a particular case of a family of numbers known as **Multinacci**. For completeness sake we report **Tribonacci** and **Pinacci numbers**, defined respectively by the recurrences

$$t_n = t_{n-3} + t_{n-2} + t_{n-1}, \qquad t_0 = 0, \qquad t_{r \leq 3} = 1, \quad (3.5.7)$$

and

$$m_n = \sum_{j=0}^{p} q_{n-j}, \qquad q_0 = 0, \qquad q_{r \leq j} = 1. \qquad (3.5.8)$$

We have used this intermezzo as a break, we have just entered the dangerous lands of pseudo-scientific thoughts and we have been shocked by a very risky hypothesis.

It is time to return to the safe lands of computation.

3.6 Series Expansion and Square Roots

Other methods, of more modern nature, allow approximations of the square roots. They are associated with techniques of calculus, like the well known **Newton binomial expansion** (NBE).

Table 3.1: **Sun-Planets Distances**

Planet	Effective Distance	Bode	Tetranacci
Mercury	0.39	0.40	0.40
Venus	0.72	0.70	0.70
Earth	1.00	1.00	1.00
Mars	1.52	1.60	1.60
Asteroids	2.70	2.80	2.80
Jupiter	5.20	5.20	5.20
Saturn	9.64	10.00	9.70
Uranus	19.20	19.60	18.40
Neptune	30.10	38.80	35.20
Pluto	39.40	77.20	67.60

The NBE of the function $f(x) = \sqrt{x^2 + 1}$ reads[14]

$$f(x) \simeq 1 + \frac{1}{2}x^2 - \frac{1}{8}x^4 + \frac{1}{16}x^6 - \frac{5}{128}x^8 + \frac{7}{256}x^{10} \cdots \qquad (3.6.1)$$

We know that the above expansion converges (namely becomes closer to the exact value) with the increasing order of the expansion, for values of the variable such that $|x| \ll 1$, while outside this interval it starts to diverge. The binomial and continued fraction expansions are therefore complementary, as shown in Fig. 3.9.

We can apply the NBE outside the convergence interval by employing the following trick, which yields a further approximation of

---

[14] Any function $g(x) = \sum_{r=0}^{\infty} \frac{g^{(n)}(x_0)}{n!}(x-x_0)^n$ may be expanded in series around a point x_0, where $g(n)(x_0)$ is the derivative of order n calculated in x_0. The series represents the function itself if it converges for all values of x, if the convergence is limited to a region around x_0 few terms of the series may provide a local approximation of the function itself.

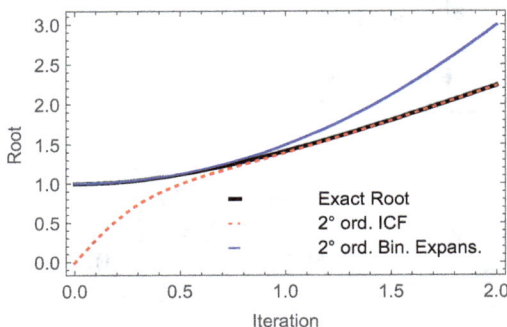

Figure 3.9: Comparison through binomial and continued fraction expansions and exact root of $\sqrt{1 + x^2}$.

our problem

$$R = p\sqrt{1 + \left(\frac{1}{p}\right)^2} \simeq p\left(1 + \frac{1}{2p^2} - \frac{1}{8p^4}\right) = p + \frac{1}{2p} - \frac{1}{8p^3} = \frac{8p^4 + 4p^2 - 1}{8p^3}.$$
(3.6.2)

We propose here a few exercises useful to provide some information of practical nature and to test the level of acquired confidence in the topics discussed so far.

Exercise 7. *Show that, for $m > p$, the square root of the sum of two squares can be approximated as (use the second order of ICF)*

$$\sqrt{m^2 + p^2} = m\sqrt{1 + \left(\frac{p}{m}\right)^2} \simeq \frac{4p^3 + 2mp^2 + 3m^2p + m^3}{4p^2 + 2mp + m^2}. \quad (3.6.3)$$

Make a comparison with the NBE method.

Exercise 8. *Use the previous result to prove the following "quasi rational" approximation*

$$\sqrt{1 + Q} = \sqrt{1 + \left(\sqrt{Q}\right)^2} \simeq \frac{4\sqrt{Q^3} + 2Q + 3\sqrt{Q} + 1}{4Q + 2\sqrt{Q} + 1}, \quad (3.6.4)$$

which can be improved if a rational approximation of $\sqrt{Q}$ is known.

Exercise 9. *Prove that any second degree equation admits a solution in terms of ICF.*

(Hint: Consider the equation

$$Ax^2 = Bx + C, \tag{3.6.5}$$

reduce it to

$$\xi^2 = b\xi + 1, \qquad \xi = \sqrt{\frac{A}{C}}\, x, \qquad b = \frac{B}{C}\sqrt{\frac{C}{A}} \tag{3.6.6}$$

and finally end up with

$$\xi = \overline{[b]} \;\rightarrow\; x = \sqrt{\frac{C}{A}}\left[1, \overline{\frac{B}{C}\sqrt{\frac{C}{A}}}\right] .) \tag{3.6.7}$$

After having dealt with square roots and second degree equations it is perhaps worthwhile to go back to the problem of cubic roots. The forthcoming section is aimed at providing further computational tools in this direction.

3.7 Cubic Roots

In the introductory section of this chapter we have quoted the immoral nature of a procedure which, even though contemptible, has allowed the definition of QI in terms of continued fractions.

In this section we use, an even more promiscuous procedure, to derive cubic roots. Let us therefore consider the equation

$$x^3 = a \tag{3.7.1}$$

which can be rewritten as

$$x^2(1+x) = a + x^2 \tag{3.7.2}$$

which yields

$$x = \sqrt{\frac{a + x^2}{1 + x}}. \tag{3.7.3}$$

The following recursion can be envisaged

$$x_0 = k, \qquad x_{n+1} = \sqrt{\frac{a + x_n^2}{1 + x_n}} \qquad (3.7.4)$$

where x_0 is an initial trial value. The last iteration is popularly known as the **"Sirventese Algorithm"**[15].

In Fig. 3.10a we have shown the algorithm applied to the derivation of $\sqrt[3]{7894}$ after choosing the trial value $k = 0$, a reasonable approximation (namely a relative error below 0.1% is obtained after few iterations ($n = 5, 6$), pushing the iteration the approximation becomes much better as shown in Fig. 3.10b where we have reported the relative error.

Given the wide opportunities offered by the uncertainties of the prescription different flavors of the "Sirventese algorithm" can be produced. The extension to higher order roots is almost straightforward and we can easily infer that the extraction of the fourth root of a given number can be reduced to the evaluation of a third order root, namely

$$x_0 = k, \qquad x_{n+1} = \sqrt[3]{\frac{a + x_n^3}{1 + x_n^2}} \qquad (3.7.5)$$

while, regarding the mth root we find

$$x_0 = k, \qquad x_{n+1} = \sqrt[m-1]{\frac{a + x_n^{m-1}}{1 + x_n^{m-2}}}. \qquad (3.7.6)$$

---

[15]The algorithm traces back to an anonymous manuscript of the 12th century which reported the method written in "Rime Sirventesi" which can be translated in English as:

A number is beautiful if it is enchanting
If its root is reduced and a number is produced
One degree is lowered and a cube is revealed.

That's all, but be cautious the historical source is not completely reliable.

Even though reasonable the procedure is not extremely practical. We have not produced a rational approximation, which can be achieved through different means we have already touched on. We consider indeed the following reshuffling of the cubic root identity

$$3x^3 - 2x^3 = a, \tag{3.7.7}$$

which can be exploited to write the recurrence

$$x_0 = k, \qquad x_{n+1} = \frac{1}{3}\left(2x_n + \frac{a}{x_n^2}\right). \tag{3.7.8}$$

We have, accordingly, ended up with a rational approximation, which leads to the iteration reported in Fig. 3.11a, along with associated relative error in Fig. 3.11b. The rational approximation is evidently much better than the previously quoted schemes in terms of speed, namely low number of steps to reach good approximation levels.

To make more evident the previous statement we report in Fig. 3.11 the derivation of the cubic root of 7894 by using the rational approximation (3.7.8) which shows its significant advantage in terms of the achieved precision. We have already noted that this last procedure is nothing but a particular case of the Newton method (also known as the method of the tangents).

The relevant use to infer a rational approximation of square roots of higher order is easily obtained and written as

$$x_0 = k, \qquad x_{n+1} = \frac{1}{m}(m-1)x_n + \frac{a}{x_n^{m-1}}. \tag{3.7.9}$$

In the forthcoming chapters we will further dwell on problems of this nature. Before closing this section it is worthwhile to invite the reader to be extremely wary towards recursive schemes. We can generate the algorithm of extraction of a cubic root through the following identity

$$x(1+x^2) = a + x \tag{3.7.10}$$

amenable to the recursive scheme

$$x_0 = k, \qquad x_{n+1} = \frac{a + x_n}{1 + x_n^2} \tag{3.7.11}$$

which does not work!!!

It can be easily checked that the method of successive iterations does not provide any reliable conclusion. The algorithm is said to be unstable. This is a technical problem which we will not discuss here.

How can we conclude this section? We have gone a little further on the Heron algorithm, we have mixed old and (relatively) new. We have learned a little bit more, but we have (fortunately) a long way to go.

3.8 What about Transcendental Numbers and Continued Fractions?

We have learned how to frame the irrational numbers within the context of CF and we have not yet mentioned *transcendental numbers* (TN) and their expansions in terms of ICF. Before getting into this aspect of the problem (just touched here in qualitative terms only), let us remind what TN are.

Transcendental numbers are "worst" than irrationals, being either not expressible as a ratio between integers and not algebraic, being not solutions of an algebraic equations with rational coefficients. An appropriate frame to fence them in is the family of *non algebraic-irrationals*. Unlike prime numbers, nobly solitary and rarefied, TN tend to stay close to each other in a promiscuous context. As shown by to Cantor, there is an uncountable number of TN in any interval (a, b). Notwithstanding recognizing one of them is absolutely not an easy task.

Liouville showed that the number

$$L = \sum_{k=1}^{\infty} 10^{-k!} \tag{3.8.1}$$

is transcendental. The transcendental nature of pi was supposed by Lambert, who proved the relevant irrationality; Hermite proved that the Napier number e is transcendental by the use of a technique later exploited by Lindemann to prove the transcendence of π.

Finally **Gelfond**, solving the 10th Hilbert problems, showed that numbers of the type a^b are transcendental if a is an algebraic (different from $0, 1$) and b is an irrational algebraic.

The reluctance of these numbers to declare their nature is well known. The **Euler Mascheroni constant**

$$\gamma = \lim_{n \to \infty} \left(\sum_{r=1}^{n} - \ln(n) \right)$$

$$= 0.5772156649015328606065120900824024310421593359 3992$$
$$(3.8.2)$$

and the **Apery constant**

$$\zeta(3) = \sum_{n=1}^{\infty} \frac{1}{n^3}$$

$$= 1.2020569031595942853997381615114499907649862 92$$
$$(3.8.3)$$

did not come out yet.

There are rational approximation of π in terms of LCF

$$\frac{22}{7} = \frac{1}{3 + \frac{1}{7}} = [3; 7] = 3.14286 \simeq \pi,$$

$$\frac{333}{106} = \frac{1}{3 + \frac{1}{7 + \frac{1}{15}}} = [3; 7, 15] = 3.14151 \simeq \pi,$$

$$\frac{355}{113} = \frac{1}{3 + \frac{1}{7 + \frac{1}{15+1}}} = [3; 7, 15, 1] = 3.14159 \simeq \pi,$$

$$(3.8.4)$$

which are mere truncated expression of the decimal expansion

$$\pi \simeq 3.14259265 = [3; 7, 15, 1, 292, 1, 1, 1, 2, 1, 3, 1, 14, 2, 1, 1, \ldots]$$
$$(3.8.5)$$

An actual expansion of π in ICF is something different as we will see below and, in general is not an easy task. Our account is just informative and elementary and is aimed just at completing the so far attempted "scenic view".

Let us outline a strategy to accomplish this task, by closely following the procedure suggested by **Euler** in his essay (see footnotes [16] and [17]):

"*De Transformatione Serium in Fractiones Continuas Ubi Simul Haec Theoria Non Mediocriter Amplificatur*"

We refer to the original paper because it is by far the most clear account on the subject. The steps we will follow are reported below:

a) We use the so far adopted notation reported below for comodity

$$[a_0;\ b_1, a_1;\ b_2, a_2; \ldots;\ b_n, a_n; \ldots] = a_0 + \cfrac{b_1}{a_1 + \cfrac{b_2}{a_2 + \cfrac{b_3}{a_3 + \cdots}}}$$

(3.8.6)

to indicate a general form of ICF;

b) According to Euler, the following Theorem holds: "*Having the series*

$$s = \frac{1}{\alpha} - \frac{1}{\beta} + \frac{1}{\gamma} - \frac{1}{\delta} \cdots$$

(3.8.7)

then the following ICF can be established

$$s^{-1} = \lfloor \alpha;\ \alpha^2, \beta - \alpha;\ \beta^2, \gamma - \beta, \ldots \rfloor.$$"

(3.8.8)

We will not provide the (not straightforward) proof and note that it can be checked by iteration. A fairly immediate consequence is given

---

[16] Commentatio 593 indicis Enestroemiani Opuscula analytica 2, 1785, p. 138–177.

[17] The Latin title can translate as "About the transformations of series into continued fractions, both theories are here expanded in detail". For the translation of the essay see also http://people.math.osu.edu/sinnott.1/ReadingClassics/continuedfractions.pdf .

below. It is well known that

$$\frac{\pi}{4} = 1 - \frac{1}{3} + \frac{1}{5} - \frac{1}{7} + \cdots \tag{3.8.9}$$

which evidently yields the ICF

$$\frac{4}{\pi} = [1; 1, 2; 9, 2; 25, 2; \ldots]. \tag{3.8.10}$$

A further almost free result is

$$\log(2) = \sum_{r=1}^{\infty} \frac{(-1)^{r+1}}{r} \tag{3.8.11}$$

which eventually yields

$$\frac{1}{\log(2)} = [1; 4, 1; 9, 1; 16, 1; \ldots]. \tag{3.8.12}$$

The Euler method can also be applied to get an ICF expansion of the Napier number, which is specified by the following series[18]

$$e^{-1} = \sum_{r=0}^{\infty} \frac{(-1)^r}{r!} \tag{3.8.13}$$

and, according to the previously outlined procedure, we find

$$e = [2; 2 \cdot 2!, 2!^2; 3 \cdot 3!, 3!^2; 4 \cdot 4!, 4!^2; \ldots] \tag{3.8.14}$$

which does not appear particularly elegant as the equivalent form proposed by Euler[19]

$$e = [2; 1, 2, 1, 1, 4, 1, 1, 6, 1, 1, 8, \ldots] \tag{3.8.15}$$

We have just provided a flavor of the interplay between TN's and ICF. The procedure we have adopted does not strictly rely upon the transcendental nature of the number under study, but on finding

---

[18]The number e is the Napier constant and is discussed in the forthcoming chapter. For the moment we consider it as the number (transcendental) to which a series of the type (3.8.13) converges.

[19]See *OEIS A003417*.

a suitably associated series.

Before concluding this section and this chapter, we quote the following expansion

$$\pi = [0; 4, 1; 1, 3; 4, 5; 9, 7; 16, 9; 25, 11; \ldots] \qquad (3.8.16)$$

in which the following pattern can be recognized

$$\pi = [0; 4, 1; \ldots n^2, 2n + 1, \ldots], \qquad n = 1, 2, \ldots \qquad (3.8.17)$$

It is not fast converging but, notwithstanding, it is remarkably beautiful.

Let us now propose the following ICF yielding an unknown number

$$? = [2; 3, 5; 7, 11; 13, 17; \ldots] \qquad (3.8.18)$$

where the a, b are arranged to be alternating prime numbers[20]. Do you believe that "?" is a TN? If so, comment the statement that it should be *more transcendental* than π.

A further element for interesting musings, with respect to the last sentence, is offered by the expansion

$$?? = [1; 2, 3, 4, 5, 6, \ldots] \qquad (3.8.19)$$

Even though the concept of continued fraction can be traced back to the work of Rafael Bombelli (1526 − 1572) we moved away from the Vedic math, but only to whet the appetite.

---

[20]http://www.herkommer.org/misc/contfrac.htm.

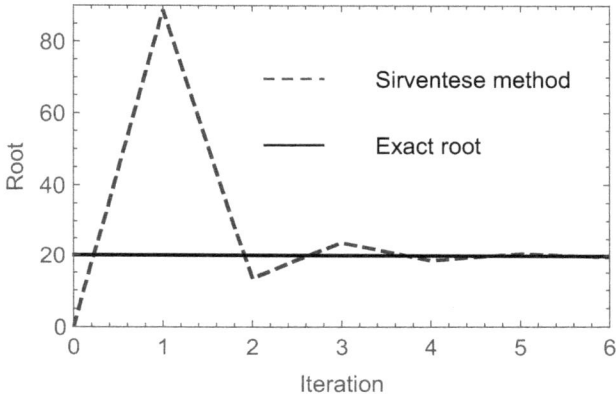

(a) Comparison between Sirventese method and exact root.

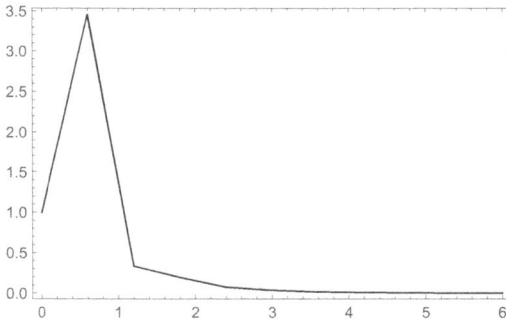

(b) Relative error %.

Figure 3.10: Sirventese calculus for $\sqrt[3]{7894}$.

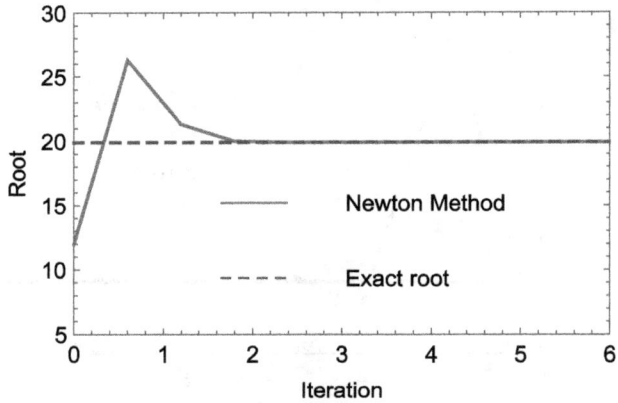

(a) Comparison between Newton method and exact root.

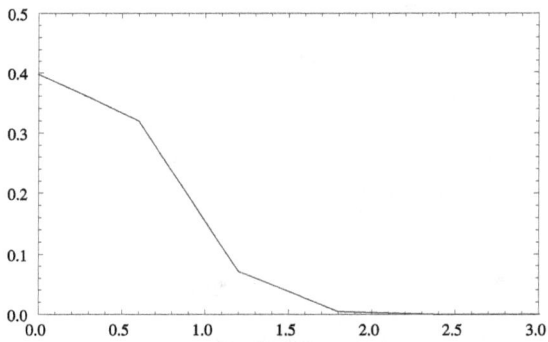

(b) Relative error %.

Figure 3.11: Newton method for $\sqrt[3]{7894}$ and $x_0 = 12$.

Chapter 4

"Imaginary Numbers", Just an Unhappy Way of Saying

4.1 Napier Number and Associated Functions

In this chapter we introduce different forms of "Imaginary" numbers or better of generalized kinds of complex numbers. We will privilege a geometrical, rather than algebraic, characterization and to this aim we need just a few remarks concerning the **Napier number** e, defined as

$$e = \lim_{n \to \infty} \left(1 + \frac{1}{n}\right)^n = 2.71828\ldots . \tag{4.1.1}$$

We can understand the genesis of this number (yielding the basis of natural logarithms) starting from the familiar concept of capitalization. It is well known that if one invests a capital C at a yearly interest rate r, after n years one will find an increment of C corresponding to

$$C_n = (1 + r)^n C \tag{4.1.2}$$

which is the formula of composite interest. Accordingly, after a year, one will find $C_1 = (1 + r)C$ and the "gain" will be $r \cdot C$. Let us now consider a daily interest of $\frac{r}{365}$ and ask ourselves the question of what is the gain after one year (365 days). We might be attempted to say that it is the same amount, the answer is not correct and the reason is evident. The use of the composite interest formula, applied to the daily interest rate yields, after one year, the capital raised to

$$\left(1 + \frac{r}{365}\right)^{365} C \neq (1 + r)C. \tag{4.1.3}$$

In order to get a feeling of how much naïve and correct answer will differ, we note that, since

$$\lim_{n \to \infty} \left(1 + \frac{\alpha}{n}\right)^n = e^\alpha \tag{4.1.4}$$

and being formula[1] $e_{365}(r) = \left(1 + \frac{r}{365}\right)^{365} \simeq e$, we can conclude that after one year the amount of the invested capital has become $e^r \cdot C$, which deviates from $(1 + r)C$ the higher the rate r is. If r were 100% instead of getting $2C$ we get $2.718C$.

The exponential function

$$Y(x) = e^x \tag{4.1.5}$$

plays a privileged role within the realm of functions, being an "eigen function" of the derivative operator, namely

$$\frac{d}{dx}y(\lambda\, x) = \lambda\, y(\lambda\, x). \tag{4.1.6}$$

A trivial consequence of this property is the derivation of the relevant series expansion

$$e^{(\lambda\, x)} = \sum_{n=0}^{\infty} \frac{(\lambda\, x)^n}{n!} \tag{4.1.7}$$

and its link with the circular functions when $\lambda = i$, where i is the imaginary unit, characterized by the following algebraic rule

$$i^2 = -1 \tag{4.1.8}$$

which, once iterated $\forall n \in \mathbb{N}$, yields

$$i^3 = -i, \quad \dots \quad , \quad i^{4n} = 1, \dots \tag{4.1.9}$$

---

[1]It should be noted that the relative error $\frac{e - e_{365}(r)}{e} = 1.366 \cdot 10^{-3}$.

Furthermore, since according to the first condition the unit i can be viewed as the solution of the second degree equation $x^2 = -1$, the second solution $-i$ will be said the (complex) conjugate.

The argument proving the link of e^{ix} to the circular functions, goes as it follows:

a) Define the function $F(x) = \cos(x) + i \sin(x)$, note that $F(0) = 1$;

b) Keep its derivative with respect to x and get
$F(x) = -\sin(x) + i \cos(x)$;

c) Note that $F(x) = i\, F(x)$. This is a first order differential equation of the type

$$\begin{cases} Y(x) = i\, Y(x) \\ Y(0) = 1 \end{cases} \qquad (4.1.10)$$

which is easily integrated and yields

$$F(x) = \cos(x) + i \sin(x) = e^{ix} \qquad (4.1.11)$$

Accordingly, by noting that the conjugate of i is $-i$ and that of $F(x)$ is $\cos(x) - i \sin(x) = e^{-ix}$, we get the **Euler formulae** for sin and cos functions as

$$\cos(x) = \frac{e^{ix} + e^{-ix}}{2}, \qquad (4.1.12)$$

$$\sin(x) = \frac{e^{ix} - e^{-ix}}{2i}. \qquad (4.1.13)$$

The relevant series expansions follows from that of the e^x and write

$$\cos(x) = \sum_{n=0}^{\infty} \frac{(-1)^n}{(2n)!} x^{2n}, \qquad (4.1.14)$$

$$\sin(x) = \sum_{n=0}^{\infty} \frac{(-1)^n}{(2n+1)!} x^{2n+1}. \qquad (4.1.15)$$

It is worth stressing that an analogous procedure allows to introduce the so called *hyperbolic functions* which as it is well known satisfy the properties ($\frac{d}{dx} = D_x$)

$$D_x(\cosh(x)) = \sinh(x), \tag{4.1.16}$$
$$D_x(\sinh(x)) = \cosh(x). \tag{4.1.17}$$

Accordingly we find that if such functions do exist with the conditions $\cosh(0) = 1$, $\sinh(0) = 0$, then

$$\cosh(x) + \sinh(x) = e^x \tag{4.1.18}$$

and

$$\cosh(x) = \frac{e^x + e^{-x}}{2} = \sum_{n=0}^{\infty} \frac{1}{(2n)!} x^{2n}, \tag{4.1.19}$$

$$\sinh(x) = \frac{e^x - e^{-x}}{2} = \sum_{n=0}^{\infty} \frac{1}{(2n+1)!} x^{2n+1}. \tag{4.1.20}$$

In the following sections we will see how profound is the relationship between circular and hyperbolic functions, if viewed within the framework of generalized "imaginary" numbers.

Before going further, we like to go back to the Pythagorean triples, which can be viewed within the framework of imaginary numbers and Euler type formulae, as reported below.

We have defined a Pythagorean triple as three integers Q, R, S satisfying

$$Q^2 + R^2 = S^2. \tag{4.1.21}$$

It was known since the Euclid times that it can be parameterized in terms of two integers[2] p, q such that

$$Q = p^2 - q^2, \qquad R = 2pq, \qquad S = p^2 + q^2. \tag{4.1.22}$$

---

[2]Euclid, "Elements", 300 B.C. Proposition 29, Lemma 1. See also K. Ryde, Trees of Primitive Pythagorean Triples, http://user42.tuxfamily.org/triples/index.html.

The two triples can be geometrically arranged as reported in Fig. 4.1 in which it is easily checked that

$$p = \sqrt{\frac{S+Q}{2}} = \sqrt{S}\cos\left(\frac{\theta}{2}\right), \quad q = \sqrt{\frac{S-Q}{2}} = \sqrt{S}\sin\left(\frac{\theta}{2}\right), \quad \alpha = \frac{\theta}{2}.$$

(4.1.23)

The use of the complex notation yields

$$p + iq = \sqrt{S}e^{i\frac{\theta}{2}}$$

(4.1.24)

which can be exploited to write

$$Se^{i\theta} = (p+iq)^2 = p^2 - q^2 + 2ipq.$$

(4.1.25)

After comparing real and imaginary parts, the previous identity yields a transparent geometrical interpretation of the identities in eqs. (4.1.22).

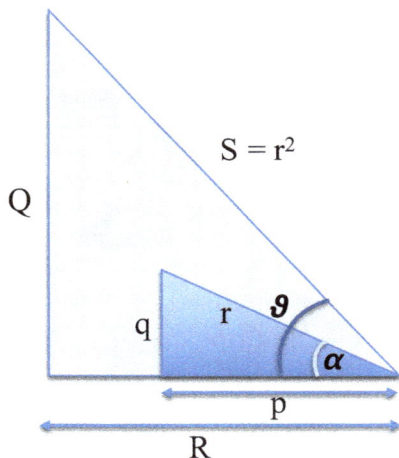

Figure 4.1: Geometry of Pythagorean triple and of the Euclidean parameterization.

4.2   Imaginary and Complex Numbers, Just an Unhappy Way of Saying

In the previous chapter we have considered the identity

$$x^2 = a + bx \tag{4.2.1}$$

and we have seen that an entire new world can emerge from its solutions if expressed in terms of continued fractions, nested radicals or quadratic irrationals.

We can make a step further in this direction by assigning to the "unknown" x the status of an imaginary number, which will be called Q-number. The relevant properties in terms of algebraic recurrences are easily derived.

Multiplying indeed both sides of eq. (4.2.1) by x we find

$$x^3 = ax + bx^2 = ax + b(a+bx) = a_1 + b_1 x \Rightarrow \begin{cases} a_1 = ab \\ b_1 = a + b^2 \end{cases} \tag{4.2.2}$$

which can be extended by iteration thus finding that the n-th power of x can be written as

$$x^n = a_{n-2} + b_{n-2} x. \tag{4.2.3}$$

The coefficients a_n, b_n are obtained by iteration according to the procedure outlined below. Multiplying eq. (4.2.3) by x we find

$$x^3 = a_0 x + b_0 x^2 = a_1 + b_1 x, \qquad b_1 = a_0 + b_0^2, \qquad a_1 = a_0 b_0, \tag{4.2.4}$$

repeating the process, we obtain

$$x^4 = a_2 + b_2 x, \qquad b_2 = b_1 b_0 + a_1, \qquad a_2 = b_1 b_0, \tag{4.2.5}$$

thus eventually ending up with the following recurrences for the a_n, b_n coefficients

$$b_n = b_{n-1} b_0 + a_{n-1}, \qquad a_n = b_{n-1} a_0. \tag{4.2.6}$$

What are the consequences we can draw from the previous statements? The iteration process is only more computationally cumbersome than the method we have followed to infer the algebraic properties of i. It is indeed evident that by setting $a = -1$, $b = 0$ in eq. (4.2.1) we get from (4.2.3) the cyclic properties of the integer powers of the ordinary imaginary unit. The unknown x can therefore be viewed as a generalized complex number. On the basis of this point of view the rational number $\frac{5}{2}$, satisfying the identity

$$\left(\frac{5}{2}\right)^2 = a + b\left(\frac{5}{2}\right), \qquad a = \frac{5}{4}, \qquad b = 2, \qquad (4.2.7)$$

is not less imaginary than i. We can therefore conclude that any number satisfying a second degree algebraic equation can be viewed as equivalent to a kind of imaginary unit.

According to this statement, we can conclude that all numbers are imaginary.

Any number Q is defined along with its conjugate according to

$$Q_+(a, b) = \frac{b + \sqrt{b^2 + 4a}}{2}, \qquad Q_-(a, b) = \frac{b - \sqrt{b^2 + 4a}}{2}. \tag{4.2.8}$$

If the symbol "$\cdot$" denotes the operation of conjugation, we find

$$\overline{(Q_+(a, b))} = Q_-(a, b), \qquad Q_+(a, b)Q_-(a, b) = -a,$$
$$Q_+(a, b) + Q_-(a, b) = b, \qquad Q_+(a, b) - Q_-(a, b) = \sqrt{b^2 + 4a}. \tag{4.2.9}$$

These Q-numbers are more general than the quadratic irrational, discussed in the previous chapter and, for example, the golden ratio is a Q-number specified by

$$\varphi = Q_+(1, 1). \tag{4.2.10}$$

The ordinary complex have a well-defined geometrical meaning, associated with rotations in a plane. The successive powers of i can be adjusted in a plane, in which on the abscissa are reported the real numbers while on the ordinate the imaginary quantities (see Fig. 4.2), according to the following rule expressed in form of multiplication table

$$1 \to (1,0), \qquad\qquad i \cdot 1 \to (0,1), \qquad\qquad i \cdot i \to (-1,0),$$
$$i \cdot i \cdot i \to (0,-1), \qquad i \cdot i \cdot i \cdot i \to (1,0), \qquad\qquad \ldots$$

$$(4.2.11)$$

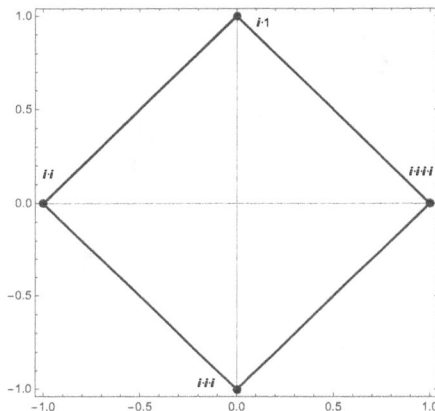

Figure 4.2: Geometrical representation of n-th powers of i.

We can define a Q-analog of the complex plane, for e.g. the golden number, according to the multiplication table

$$1 \to (1,0), \qquad\qquad \varphi \cdot 1 \to (0,\varphi), \qquad\qquad \varphi \cdot \varphi \to (1,\varphi),$$
$$\varphi \cdot \varphi \cdot \varphi \to (1,2\varphi), \quad \varphi^4 = \varphi^2 \varphi^2 \to (2,3\varphi), \quad \varphi^5 = (3,5\varphi), \quad \ldots$$
$$\varphi^n \to (F_{n-1}, F_n \varphi)$$

$$(4.2.12)$$

where F_n are the already quoted Fibonacci numbers.

The behavior is not cyclical, as in the ordinary case, but that of a broken line (see later Fig. 4.3), approaching the line with equation

$$y = \lim_{n\to\infty} \frac{F_{n+1}}{F_n} \varphi x = \varphi^2 x \qquad (4.2.13)$$

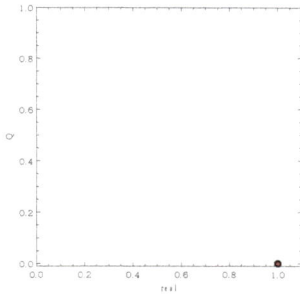

(a) $\varphi^0 = \{1, 0\}$.

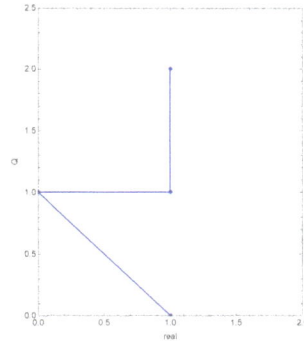

(b) $\varphi^n : n = 0, 1, 2, 3$.

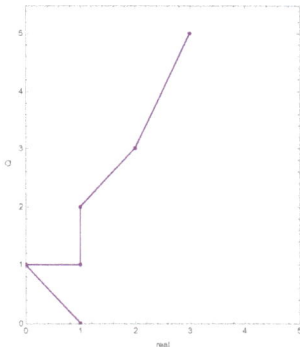

(c) $\varphi^n : n = 0, 1, \ldots, 5$.

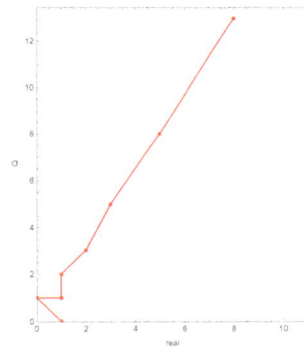

(d) $\varphi^n : n = 0, 1, \ldots, 7$.

Figure 4.3: Geometrical representation of n-th powers of numbers associated with Fibonacci numbers.

A similar behavior can be obtained for Q-numbers associated with the silver ratio, as reported later in Fig. 4.4.

In less naïve terms we should say that the imaginary unit generates an *Algebraic structure* while Q does not satisfy this property.

4.3 Matrices as Multidimensional Numbers

Before starting, we like to underscore that we felt uneasy when we compiled the multiplication table of imaginary numbers. There

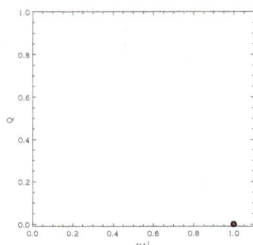

(a) $\sigma^0 = \{1, 0\}$.

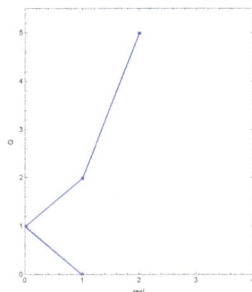

(b) $\sigma^n : n = 0, 1, 2, 3$.

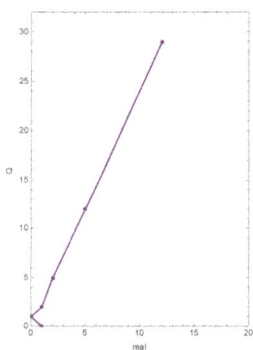

(c) $\sigma^n : n = 0, 1, \ldots, 5$.

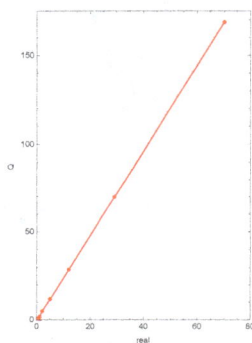

(d) $\sigma^n : n = 0, 1, \ldots, 7$.

Figure 4.4: Geometrical representation of n-th powers of numbers associated with silver ratio.

is a kind of formal incoherence deriving from the fact that we introduced the imaginary unit viewed as an operator acting on a couple of numbers, representing points on a plane. In order to define a less foggy procedure it is convenient to acquire a more suitable tool. To this aim we introduce the 2×2 matrix

$$\hat{i} = \begin{pmatrix} 0 & -1 \\ 1 & 0 \end{pmatrix} \tag{4.3.1}$$

which can be viewed as an imaginary unit. Its square yields indeed the unit matrix 1 with changed sign

$$\hat{i} = \hat{1}, \qquad\qquad \hat{1} = \begin{pmatrix} 1 & 0 \\ 0 & 1 \end{pmatrix}. \tag{4.3.2}$$

If we furthermore arrange the definition of a couple of points in the complex plane as

$$\underline{n} = \begin{pmatrix} -y \\ x \end{pmatrix} \tag{4.3.3}$$

we have solved the problem of having a formal definition of the rules underlying the definition of the multiplication table, namely

$$\hat{i}\,\underline{n} = \begin{pmatrix} -y \\ x \end{pmatrix}. \tag{4.3.4}$$

The successive actions of the imaginary unit matrix on the vector $(1,0)$ yields therefore

$$\hat{i} \cdot \begin{pmatrix} 1 \\ 0 \end{pmatrix} = \begin{pmatrix} 0 \\ 1 \end{pmatrix}, \qquad \hat{i}^2 \cdot \begin{pmatrix} 1 \\ 0 \end{pmatrix} = \begin{pmatrix} -1 \\ 0 \end{pmatrix},$$

$$\hat{i}^3 \cdot \begin{pmatrix} 1 \\ 0 \end{pmatrix} = \begin{pmatrix} 0 \\ -1 \end{pmatrix}, \qquad \hat{i}^4 \cdot \begin{pmatrix} 1 \\ 0 \end{pmatrix} = \begin{pmatrix} 1 \\ 0 \end{pmatrix}, \tag{4.3.5}$$

which represents a well-founded mathematical element supporting the geometric interpretation reported in Fig. 4.3.

We make a step further and assume that any Q-number can be represented by a matrix of the type

$$\hat{q}(a,b) = \begin{pmatrix} 0 & a \\ 1 & b \end{pmatrix} \tag{4.3.6}$$

which, as easily checked, satisfies the identity

$$(\hat{q}(a,b))^2 = a\hat{1} + b\hat{q}(a,b). \tag{4.3.7}$$

This is a clue that the previous assumption is correct. By repeating the procedure we find that the n-th power of the matrix $\hat{q}(a,b)$ can be written as

$$(\hat{q}(a,b))^n = a_{n-2}\,\hat{1} + b_{n-2}\,\hat{q}(a,b). \tag{4.3.8}$$

The properties of this family of matrices are notable and, regarding the case $\hat{q}(1,1)$, we find

$$(\hat{q}(1,1))^n = \begin{pmatrix} F_{n-1} & F_n \\ F_n & F_{n+1} \end{pmatrix}. \tag{4.3.9}$$

Regarding, e.g., the determinant we find

$$| \, (\hat{q}(a,b))^n \, | = a^n \qquad (4.3.10)$$

which can be shown to derive the following properties of Fibonacci numbers

$$| \, (\hat{q}(1,1))^n \, | = F_{n-1}F_{n+1} - F_n^2 = (-1)^n. \qquad (4.3.11)$$

An analogous result holds for the Pell numbers generated by $(\hat{q}(1,2))^n$.

We have now all the elements to understand the action of the matrix Q on a vector belonging to a Q-complex space. We consider, e.g.,

$$\begin{pmatrix} x_n \\ y_n \end{pmatrix} = (\hat{q}(a,b))^n \begin{pmatrix} 1 \\ 0 \end{pmatrix}. \qquad (4.3.12)$$

We can check that for successive n, the geometrical representation is a broken line in the Q-plane, which for large n tend to a line with inclination angle given by

$$\varepsilon = \lim_{n \to \infty} \arctan \left(\frac{b_n}{a_n} \right). \qquad (4.3.13)$$

This type of geometrical interpretation is due to Klein[3], who argued that if an irrational number α has a continued fraction expansion $[a_1, a_2, a_3, ..., a_n, ...]$ with reduced factors $c_i = \frac{p_i}{q_i}$, then forming the points defined by the couples $(q_1, p_1), (q_2, p_2), (q_3, p_3), \ldots$ we discover that the line with slope ε interpolates the points defined by (q_i, p_i).

Before going further it is worth noting that any 2×2 matrix can be written as the product of Q-matrices numbers as it follows

$$\begin{pmatrix} a & b \\ c & d \end{pmatrix} = \hat{q}(a,c)\hat{q}\left(\frac{ad-bc}{a}, \frac{b}{a} \right) \qquad (4.3.14)$$

whose importance will be underscored below.

---

[3]F. Klein, Augeswalthe Kapitel der Zalentheorie, Teubner, 1907, pp. 17–25.

4.4 *Q*-Trigonometry

A little abstraction is now in order if we like to go deeper in the so far developed discussion. We are therefore obliged to introduce a few notions, not of elementary nature, concerning the geometrical nature of complex numbers.

The square modulus of a complex number is defined as the product of the number itself times its complex conjugate. The geometrical interpretation of the modulus is the length of the vector associated to the complex number in the **Argand-Gauss plane**. A *Q*-complex number and its conjugate are specified by

$$Z = X + Q_+Y, \qquad\qquad Z^* = X + Q_-Y. \qquad (4.4.1)$$

Furthermore, we define the *Q*-real and *Q*-imaginary parts of Z according to

$$_QRe(Z) = \frac{Q_+Z^* - Q_-Z}{2\sqrt{\Delta}} = X, \quad _QIm(Z) = \frac{Z - Z^*}{2\sqrt{\Delta}} = Y \quad (4.4.2)$$

where $\Delta = b^2 + 4a$. The modulus of a *Q*-complex number is denoted by $_Q|\,Z\,|$ according to the identity

$$_Q|\,Z\,|^2 = ZZ^* = X^2 + bXY - aY^2, \qquad\qquad \forall a, b \in \mathbb{R} \qquad (4.4.3)$$

while that associated with the sum of two *Q*-complex numbers is below

$$_Q|\,Z_1 + Z_2\,|^2 = {}_Q|\,Z_1\,|^2 + (Z_1Z_2^* + Z_1^*Z_2) + {}_Q|\,Z_1\,|^2 . \qquad (4.4.4)$$

The analogy with the ordinary case is strong, it can even be strengthened.

The ordinary Euler formulae

$$e^{i\delta} = \cos(\delta) + i\sin(\delta), \qquad\qquad e^{-i\delta} = \cos(\delta) - i\sin(\delta) \quad (4.4.5)$$

can be generalized to their *Q*-counterpart by writing

$$e^{Q+\delta} = {}_QC(\delta) + Q_+ {}_QS(\delta), \qquad\qquad e^{Q-\delta} = {}_QC(\delta) + Q_- {}_QS(\delta) \qquad (4.4.6)$$

and thus obtaining for the Q-trigonometric functions

$$_QC(\delta) = \frac{Q_+ e^{Q-\delta} - Q_- e^{Q+\delta}}{2\sqrt{\Delta}}, \qquad _QS(\delta) = \frac{e^{Q-\delta} - e^{Q+\delta}}{2\sqrt{\Delta}}.$$

$$(4.4.7)$$

In Fig. 4.5 we have shown the associated behavior in the Q-complex plane for $\delta \in \{0, n \cdot 2\pi\}$:

     i) for $a < 0$ and $b = 0$ the functions (4.4.7) reduce to the ordinary cos and sin functions,

     ii) for $b > 0$ they are different and describe in the Q-plane a sort of logarithmic spiral,

     iii) for $a > 0$ their behavior is similar to the hyperbolic functions.

The fundamental identity of the Q-trigonometric functions writes

$$\left(_QC(\delta)\right)^2 + b\left(_QC(\delta)\right)\left(_QS(\delta)\right) - a\left(_QS(\delta)\right)^2 = e^{b\delta}, \qquad (4.4.8)$$

it has been obtained by expanding the product $e^{Q+\delta} \cdot e^{Q-\delta}$. From the geometrical point of view it is not simply a circle but a more general conic.

We can further push the analogy by introducing the Q-tangent function

$$_QT(\delta) = \frac{_QS(\delta)}{_QC(\delta)} \qquad (4.4.9)$$

and by writing the corresponding Q-(sin, cos) as

$$_QC(\delta) = \frac{e^{\frac{b}{2}\delta}\, _QT(\delta)}{\sqrt{\left(_QT(\delta)\right)^2 + b\,_QT(\delta) - a}},$$

$$(4.4.10)$$

$$_QS(\delta) = \frac{e^{\frac{b}{2}\delta}}{\sqrt{\left(_QT(\delta)\right)^2 + b\,_QT(\delta) - a}}.$$

We wish to end this section by focusing our attention on a point, we believe important.

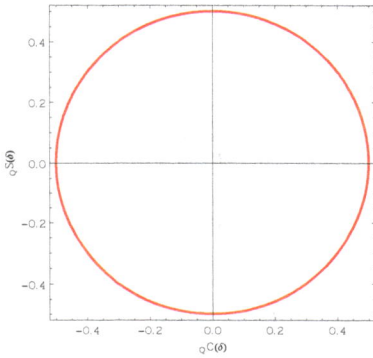

(a) $a = -1$, $b = 0$, $n = 1$.

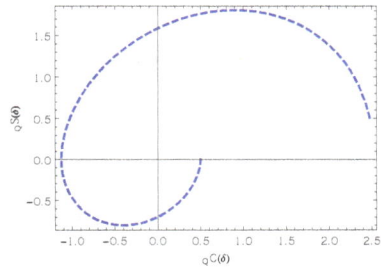

(b) $a = -1$, $b = 0.5$, $n = 1$.

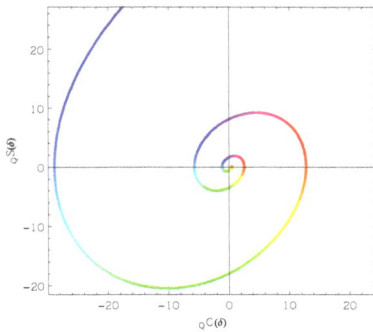

(c) $a = -1$, $b = 0.5$, $n = 3$.

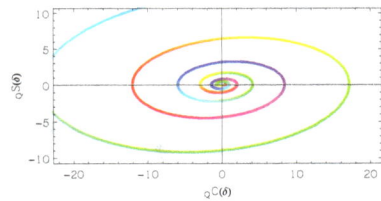

(d) $a = -5$, $b = 0.5$, $n = 3$.

Figure 4.5: Q-trigonometric representation.

If we take into account that $\hat{q}(a, b)^n = a_{n-2}\hat{1} + b_{n-2}\hat{q}(a, b)$, it is easily argued that

$$\sum_{n=0}^{\infty} \frac{\delta^n}{n!}\hat{q}(a, b)^n = \hat{1} + \delta\hat{q}(a, b) + \sum_{n=2}^{\infty} \frac{\delta^n}{n!}\left(a_{n-2}\hat{1} + b_{n-2}\hat{q}(a, b)\right) \quad (4.4.11)$$

which also yields the following series expansion for the Q-trigonometric functions

$$_QC(\delta) = 1 + \sum_{m=0}^{\infty} \frac{a_{m+2}}{(m+2)!}\delta^{m+2}, \qquad _QS(\delta) = \delta + \sum_{m=0}^{\infty} \frac{b_{m+2}}{(m+2)!}\delta^{m+1}.$$
$$(4.4.12)$$

A further point worth to be strengthened is that by defining the ordinary imaginary unit we have introduced the matrix $\hat{i} = \begin{pmatrix} 0 & -1 \\ 1 & 0 \end{pmatrix}$ which, once squared, returns minus the unit matrix. The use of matrices offer some advantages not only in terms of flexibility of notation but yields the possibility of generalizing some concepts.

If for example the matrix

$$\hat{h} = \hat{q}(1, 0) = \begin{pmatrix} 0 & 1 \\ 1 & 0 \end{pmatrix} \quad (4.4.13)$$

is used, it is natural to introduce the further imaginary unit

$$\hat{h}^2 = \hat{1}, \quad (4.4.14)$$

which will be denoted as imaginary hyperbolic unit, as a consequence of the identity

$$e^{\hat{h}\delta} = \cosh(x)\hat{1} + \hat{h}\sinh(x). \quad (4.4.15)$$

Accordingly, the hyperbolic trigonometry is a particular case of the Q-trigonometry.

4.5   So What?

It is natural to ask: "... what's the point and what is the link, if any, between this latter discussion and the matter we developed in the previous chapters, mainly in the first and the second chapters?"

Our attempt has been that of finding a common trade linking arithmetic operations, rational numbers, continued fractions, irrational numbers and a slightly more general conception regarding the complex numbers, capable of including the reduced quadratic irrationals.

An "Exoteric-Geometric" cue to get the link between what has been discussed in this chapter and the topics covered previously, is offered by the following observation.

Observation 2. *Given four Fibonacci consecutive numbers* F_n, F_{n+1}, F_{n+2}, F_{n+3}, *the following combinations*

$$F_n F_{n+3}, \qquad 2F_{n+1}F_{n+2}, \qquad F_{n+2}F_{n+3} - F_n F_{n+1} \quad (4.5.1)$$

yields a Pythagorean triple

$$(F_n F_{n+3})^2 + 4(F_{n+1}F_{n+2})^2 = (F_{n+2}F_{n+3} - F_n F_{n+1})^2. \qquad (4.5.2)$$

We have underscored that too often in divulgative papers or books it is likely to find unreliable speculations on Fibonacci numbers, which are supposed to entities with extraordinary meanings and properties. In a book, published a few years ago[4] it was declared that the golden section is the most irrational of numbers. It is evident that this statement sounds not particularly profound in absolute terms and from the mathematical point of view in particular[5]. According to the previous discussion all the quadratic irrationals are equivalent from the mathematical point of view. None of them plays any privileged role within the relevant environment.

Any succession of the type

$$G_0 = 0, \qquad G_1 = 1, \qquad G_{n+2} = yG_{n+1} + G_n \qquad (4.5.3)$$

---

[4]M. Livio, The Golden Ratio: The Story of Phi, the World's Most Astonishing Number, Broadway Books, 2002.

[5]See also C. E. Falbo, Generalizations of the Golden Ratio, http://www.mathfile.net/generalized_phi_mathpage.pdf.

is characterized by a specific "golden section" defined as

$$\varphi(y) = \frac{y + \sqrt{y^2 + 4}}{2} \qquad (4.5.4)$$

and satisfying the properties

$$(\varphi(y))^{-1} = \varphi(-y) = 1 - \varphi(y) = -(\varphi(y))^* \qquad (4.5.5)$$

which are formally interesting for their analogy with the ordinary imaginary numbers, hence the introduction of Q-trigonometry by means of the Euler generalized formulae. Furthermore if y is assumed to be an integer, a succession fully equivalent to that leading to the Fibonacci numbers will be obtained.

The use of the already quoted Binet formula allows the definition of the $G_n(y)$ numbers as

$$G_n(y) = \frac{(\varphi(y))^n + ((\varphi(y))^*)^n}{\sqrt{4 + y^2}} \qquad (4.5.6)$$

whose properties are particularly interesting to be explored. It is evident that they reduce to Fibonacci numbers for $y = 1$.

It is furthermore worth noting that

$$\sum_{n=0}^{\infty} t^n G_n(y) = \frac{1}{\sqrt{4+y^2}} \left(\frac{1}{1+t\left(y+\sqrt{4+y^2}\right)} + \frac{1}{1+t\left(y-\sqrt{4+y^2}\right)} \right)$$

$$= \frac{2}{\sqrt{4+y^2}} \left(\frac{1+yt}{1+2yt-4t^2} \right)$$

$$(4.5.7)$$

and introducing the conjugated form

$$S_n(y) = \frac{(\varphi(y))^n - (\varphi(y))^*}{\sqrt{4 + y^2}} \qquad (4.5.8)$$

we obtain

$$\sum_{n=0}^{\infty} t^n S_n(y) = \frac{1}{\sqrt{4+y^2}} \left(\frac{1}{1+t\left(y+\sqrt{4+y^2}\right)} \frac{1}{1+t\left(y-\sqrt{4+y^2}\right)} \right)$$

$$= 2 \left(\frac{1}{1+2yt-4t^2} \right).$$

$$(4.5.9)$$

The procedure we have just discussed (and will be further exploited in the following) is known as the method of the generating function, according to which we have determined a function in the t variable, providing the polynomials $S_n(y)$ e $G_n(y)$ as the coefficient of a Taylor series expansion.

The generating functions we have just obtained are associated with almost all the families of special polynomials. We like to mention a family known as Chebyshev polynomials (ohps!!! We did not choose to the Anglo-Saxon transliteration) and defined as

$$\sum_{n=0}^{\infty} t^n T_n(x) = \frac{1-tx}{1-2tx+t^2}, \qquad \sum_{n=0}^{\infty} t^n U_n(x) = \frac{1}{1-2tx+t^2}.$$

$$(4.5.10)$$

If they are compared with those relevant to the Chebyshev family and by applying the *Principle of Identity of Polynomials*[6] we get the following identities

$$G_n(y) = \frac{2^{n+1}i^n}{\sqrt{4+y^2}} T_n\left(-i\frac{y}{2}\right), \qquad S_n(y) = 2^{n+1}i^n U_n\left(-i\frac{y}{2}\right).$$

$$(4.5.11)$$

The Fibonacci numbers are therefore given by

$$F_n = \frac{2^{n+1}i^n}{\sqrt{5}} T_n\left(-i\frac{1}{2}\right).$$

$$(4.5.12)$$

---

[6]By the Identity Principle it is meant that two polynomials $p_n(x) = \sum_{r=0}^{n} a_r x^r$ and $\pi_n(x) = \sum_{s=0}^{n} b_s x^s$ are equal if and only if the coefficients corresponding to the variable x are equal, namely if $a_m = b_m$, $\forall m$.

What are we trying to convey?

There is a widespread attitude of attributing to F_n and to the golden ratio a privileged position according to which they are the key of a secret code in Physics, Biology, Anatomy, Architecture... to be deciphered in terms of "ϕ".

What we like to underscore is that F_n and ϕ and any other device of the same kind have not any exoteric meaning and all the associated speculations are mere not justified, often self-referential, speculations. The following quote is important: " ... *to document all the wrong information on the golden ratio, a very voluminous book would be needed, many of these things are the result of the same errors interpreted by different authors*".

The enthusiastic supporters of the codes inspired by ancient wisdom cite irrefutable evidences such as the one reported in the following[7] Fig. 4.6, which shows a drawing attributed to Leonardo[8]. The superimposed rectangles added by the artist for a study concerning a study of the relative proportions, are interpreted as they are golden rectangles, regardless of the relative lengths.

Now, to reconcile the discussion of this chapter with that of the previous chapters, let's consider a number $r > 1$ and we put $p = r - \frac{1}{r}$, from which follows $r^2 = rp + 1$ (if $p = 1$ we will have $r = \phi$); now we associate $rp + 1$ with what is shown in Fig. 4.7, that is the diameter BC of a semicircle, and we build the right triangle inscribed in it, in which $r * p$ and 1 are the projections of the related cathets on the hypotenuse.

The figure allows the further clarifications:

1) According to the second Euclid Theorem $\overline{BH} : \overline{AH} = \overline{AH} : \overline{CH}$ and therefore $\overline{AH} = \sqrt{rp}$ (this was the method adopted by an-

---

[7]Public imagine http://entokey.com/aesthetic-facial-analysis.

[8]G. Markowsky, "Missconceptions about the golden ratio", The college mathematical journal n. 23, 1992, pp. 2–19.

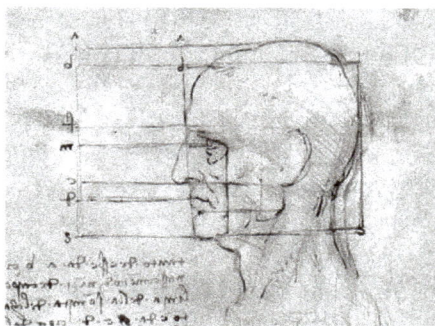

Figure 4.6: The proportions of the head, Leonardo da Vinci, ca. 1490.

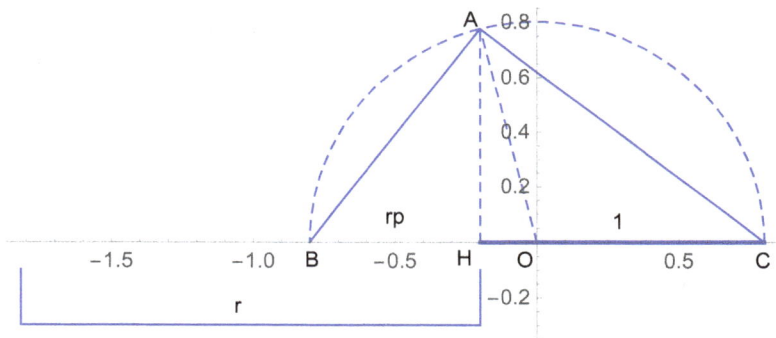

Figure 4.7: Keplero triangle construction.

cient Greeks to calculate the square roots).

2) The hypotenuse of the rectangle triangle is $\overline{BC} = rp + 1 = r^2$ and if $p = 1$ we find $\overline{BC} = \phi^2$.

In this chapter we travelled from Algebra to Geometry, using the notion of an imaginary unit, which we exploited in a wider context. This is not the end of the story. The world of polynomials and of the associated forms of generalized imaginary numbers reserves further surprises to us as we will see in the forthcoming chapter.

As a final practical notion we try to derive a "ruler" (probably used by the Greeks) for the calculation of square roots.

Example 12. *Suppose therefore we want to calculate $\sqrt{3}$ with ruler and compass. Let's take a $\overline{BC}$ segment in length $4 = 3 + 1$ and on this we mark a segment of $\overline{HC}$ of length 1 and $\overline{BH}$ of length 3. We construct the perpendicular to the segment $\overline{BC}$ in H and trace the semicircle of diameter $\overline{BC}$. A is the meeting point between the semicircle and the perpendicular $\overline{AH}$ is the square root of 3.*

We conclude with an additional (geometric) calculation tool for the determination of quantities associated with the golden ratio.

Example 13. *The use of identity $\frac{1}{\phi^2} + \frac{1}{\phi} = 1$ allows to obtain what is shown in the Fig. 4.8 where a circle of radius $\frac{1}{2}$ and the tangent parallel to the diagonal of the rectangle of sides $\frac{1}{2}$ and 1 are reported.*

4.6   Third Degree Polynomials

Before considering the problems related to polynomials and complex numbers, let's try to put our feet on the ground again by asking ourselves the problem of how to solve the equation

$$x^3 + ax^2 + bx = c. \qquad (4.6.1)$$

Since we have reached a certain level of independence of thought, we can also believe that the calculation of x is equivalent to the extraction of a generalized type of "root" of the number c, or the known term of our equation. We will denote this operation with

$$x = {}^{(3,2a,1b}\!\sqrt{c}. \qquad (4.6.2)$$

It is evident that $a = b = 0$ everything reduces to

$$^{(3,2_0,1_0}\!\sqrt{c} = \sqrt[3]{c}. \qquad (4.6.3)$$

Considering for the moment an approximate calculation, we will resort to our immoral techniques, which helped us to define the ordinary cubic root calculation algorithm. We rewrite the previous cubic

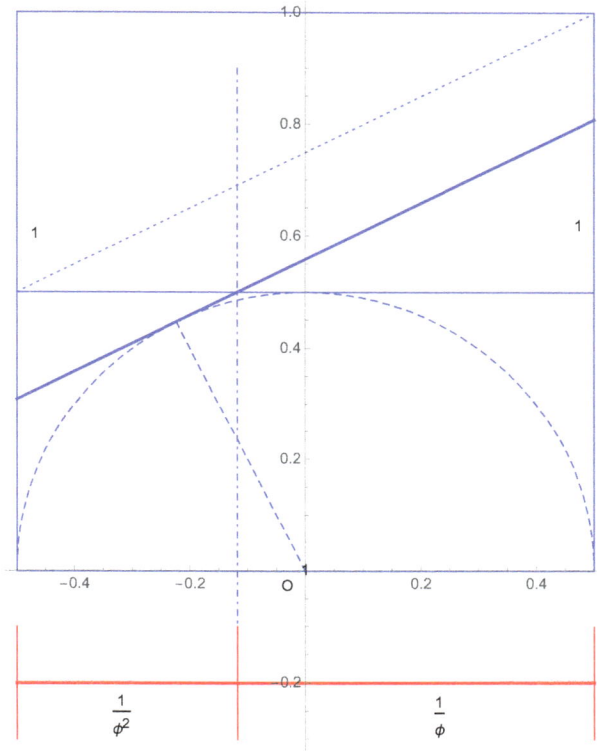

Figure 4.8: Geometrical ruler for the determination of irrational numbers $\frac{1}{\phi}$ and $\frac{1}{\phi^2}$.

equation like

$$
\begin{cases}
y = \dfrac{c + (2x - a)x^2}{b + 3x^2} \\
y = x
\end{cases}
, \tag{4.6.4}
$$

obtained after setting $by + 3x^2 y = c + 2x^3 - ax^2$. The emerging iterative algorithm is

$$
x_{n+1} = \frac{c + (2x_n - a)x_n^2}{b + 3x_n^2}, \qquad\qquad x_0 = k. \tag{4.6.5}
$$

In Fig. 4.9 we have reported an example of convergence of the procedure.

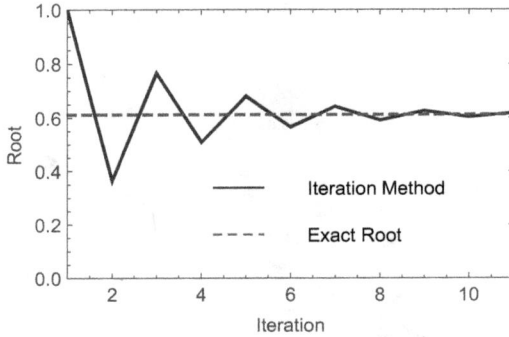

Figure 4.9: Ultraradicals $^{(3,2_5,^{18})}\!\sqrt{7}$ exact solution and iteration method with $x_0 = 1$.

Like the second degree equations, the solutions of a cubic equation are not unique and as we know are 3, linked to the coefficients of the equation by the identities

$$x_1 + x_2 + x_3 = -a, \qquad\qquad x_1 x_2 x_3 = c . \qquad (4.6.6)$$

If we indicate by y the solution provided by the iterative algorithm, we get the remaining roots provided by the second degree equation

$$X^2 + (a+y)X + \frac{c}{y} = 0 \qquad (4.6.7)$$

thus, in conclusion, finding

$$^{(3,2_a,^{1_b})}\!\sqrt{c} = \begin{cases} y \\[2mm] \dfrac{-(a+y) + \sqrt{(a+y)^2 - \frac{4c}{y}}}{2} \\[4mm] \dfrac{-(a+y) - \sqrt{(a+y)^2 - \frac{4c}{y}}}{2} \end{cases} . \qquad (4.6.8)$$

Let us now consider the fourth degree equation

$$x^4 + ax^3 + bx^2 + cx = d, \qquad (4.6.9)$$

according to the formalism of the generalized roots, the relevant solution can be written as

$$x = {}^{(4,3a,2b,1c)}\sqrt{d},$$ (4.6.10)

by applying the same procedure as before, we evaluate one of the roots with an iterative algorithm

$$x_{n+1} = \frac{d + (3x_n^2 - ax_n - b)x_n^2}{c + 4x_n^3}, \qquad x_0 = k .$$ (4.6.11)

The further roots are obtainable by operating as in the case of the third degree equation. We will say more about this generalized form of root in next chapter.

It is evident that the extension of the procedure to a higher degree equations does not present, at least from the conceptual point of view, any problem. The method we have described is (in its essential lines) a particular case of a more general technique for nonlinear equation solution known as Newton's (Raphson) algorithm.

We kept mentioning this procedure before the next chapter in which we will touch on algebraic equations of $2, 3, 4$ and higher, for two reasons.

The first is related to the logical procedure that we followed, we made the solution of the equation depend degree n from that of degree $n - 1$ which, as we shall see, was the guiding principle for the search for the formula solution of the equations of degree higher than the first.

The second is that the Babylonians had perhaps guessed the meaning of "roots" of type ${}^{(3,2a,1b)}\sqrt{c}$, in fact they had tables relating to the calculation of $n^3 + n^2$ useful for the extraction of "roots" ${}^{(3,2a,1b)}\sqrt{c}$, which shows the high level of abstraction to which the civilizations of the past had come.

Before closing, let's go back to the "complex number" defined by the relationship $x^2 = a + bx$ which, in the context of formalism suggested in this section, can be interpreted $h = {}^{(2,1-b)}\sqrt{a}$.

In the next chapters we will draw further consequences from this different point of view.

Chapter 5

Algebraic Equations: Vedic and Western Points of View

5.1   Introduction

In the previous chapter we saw how the concept of imaginary number is so intrinsically linked to the notion of algebraic equations and to their structure that the notion of imaginary number acquires a completely different meaning, becoming only an alternative definition to the same equation. The proper framework to deal with the problem would require the recourse to Galois Theory and introduce the concept of number according to this point of view.

Albeit this is the correct environment, it goes very much beyond the purposes and the interests of this discussion and such a digression would not be useful for what we have in mind. In the course of the previous chapter, we wanted to highlight how the phrase "imaginary", used to define numbers that are not real, is unhappily misleading. Given the rigor that inspires the pages of this book we completely change our mind and argue the opposite, for historical reasons and beyond.

Although the solution of the second degree equations goes back, as we have seen, to the Assyrians who discovered the method called the square complement, the notion of the imaginary number arises a few millennia later, by the Italian algebraists, engaged in the study of third degree equations.

In particular, in book I of Algebra, Rafael Bombelli introduces the square roots of the negative unit and calls these new entities "Quantitá Silvestri" loosely translated as "Sylvan Quantities" and comes to clarify the rules for operating with the numbers that come today called "complexes". For obscure reasons, at least as those relating to the denomination of sylvan quantities, distinguishes between i and $-i$, introducing the terms more than less and less than less. We have used the italian locution. It is to be noted that this is the Italian of 16th century, significantly different from the present language. In order to denote the product of the (positive) unit time the (positive) imaginary unit, Bombelli writes "*Piú via piú di meno*" which can barely be translated as "*Plus one times more than minus*", where "more than minus" indicates the imaginary unit.

Table 5.1: **Imaginary and relative unit digits multiplication table**

Locution	Modern Statement
Piú via piú di meno,fa piú di meno	$(+1) \times (+i) = +i$
Meno via piú di meno,fa meno di meno	$(-1) \times (+i) = -i$
Piú via meno di meno,fa meno di meno	$(+1) \times (-i) = -i$
Meno via meno di meno,fa piú di meno	$(-1) \times (-i) = +i$
Piú di meno via piú di meno,fa meno	$(+i) \times (+i) = -1$
Piú di meno via men di meno,fa piú	$(+i) \times (-i) = +1$
Meno di meno via piú di meno,fa piú	$(-i) \times (+i) = +1$
Meno di meno via men di meno,fa meno	$(-i) \times (-i) = -1$

The term "sylvan quantity" evokes an environment of pagan divinities, among lush woods, nymphs and Satyrs, from whose union these hybrid, Divine, Diabolic and Elusive forms were born. The context in which the above entities have sprouted is perhaps adequately illustrated in the picture below, which reproduces one idea, presumably close to the truth, of how Dionisio, Bacco, the Baccanti and the rest of the happy brigade conceived the wild numbers (Fig. 5.1)[1].

Figure 5.1: *Bacco's Youth*, H. William-Adolphe Bouguereau.

We could certainly put forward some hypotheses on the reasons that may have determined the choice to define the imaginary unit i as more than less (or *pdm*) as Bombelli wrote in his calculations, but we prefer to desist.

Let us now try to understand what solving an equation actually means and we will see what are the consequences of the assumption that equations "live" in a universe populated only by rational numbers, such as the one in the following figure (Fig. 5.2)[2].

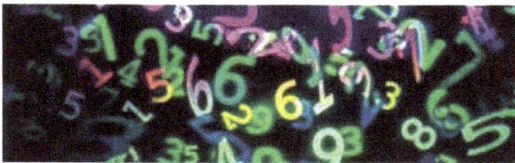

Figure 5.2: Number planet.

---

[1]From *https : //it.wikipedia.org/wiki/File : William − Adolphe_Bouguereau* $(1825 − 1905)−The\_Youth\_of\_Bacchus\_(1884).jpg$.

[2]From *https : //stefanoleonesi.wordpress.com/curriculum/*

Now suppose that the solution of our equation corresponds to the search for the perpetrator of a crime: Investigators will expect this to not escape human varieties and would certainly find themselves in difficulty if from testimonies were the evidence of a culprit with three heads, sixteen arms and blue in color.

If we are interested in solving the equation

$$x^2 - 5 = 0 \qquad (5.1.1)$$

in a universe populated only by rational numbers, we should conclude that none of the inhabitants of the planet shown in Fig. 5.2 is able to satisfy it, unless due to alien intrusion or genetic modification.

In "educated" terms we say that the set of rational numbers constitutes a $\mathbb{Q}$ *field*, or a structure closed under arithmetic operations. What we mean is that two elements of $\mathbb{Q}$ combined through the arithmetic operations always provide an element of $\mathbb{Q}$. Alien forms (or genetically modified) if they want to have legitimacy of belonging to the world they have had access must adapt to the rules that govern the universe of rational numbers, that is, they must be able to "Couple" with each other through multiplication, addition ...

For this purpose, to extend citizenship to an alien k specified by $k^2 = 5$, we define a generic element of the enlarged field $\mathbb{Q}(\sqrt{5})$ as

$$\zeta = a + \sqrt{5}b \ \in \ \mathbb{Q}(\sqrt{5}) \qquad \text{with } a, b \in \mathbb{Q}. \qquad (5.1.2)$$

The rules of composition of two distinct elements ζ_1, ζ_2 are completely analogous to those relating to the case of imaginary numbers

$$\zeta_1 \zeta_2 = m_p + \sqrt{5} \, n_p, \qquad m_p, n_p \in \mathbb{Q}$$

$$\frac{\zeta_1}{\zeta_2} = m_r + \sqrt{5} \, n_r, \qquad m_r, n_r \in \mathbb{Q},$$

$$m_p = a_1 a_2 + 5 b_1 b_2, \qquad n_p = a_1 b_2 + a_2 b_1,$$

$$m_r = \frac{a_1 a_2 - 5 b_1 b_2}{a_2^2 - 5 b_2^2}, \qquad n_r = \frac{a_2 b_1 - a_1 b_2}{a_2^2 - 5 b_2^2}, \qquad a_2^2 - 5 b_2^2 \neq 0 \, .$$

$$(5.1.3)$$

In the case of the equation $x^2 + 1 = 0$, we should accept the extension of the field of rational numbers with the introduction of a unit k such that $k^2 = -1$.

Up to now we have strengthened the point of view of the previous chapter, that is, all numbers have the same right of citizenship in a suitably enlarged field $\mathbb{Q}$ provided that these numbers respect the arithmetic rules of the field of rational numbers.

The solution of the equation $x^2 = x + 1$ it is therefore guaranteed in the extended field $\mathbb{Q}(k)$, $k^2 = k + 1$. The only problem is that there are two options k_+ and k_- linked from the relation $k_- = -(k_+)^{-1}$, in the previous chapter we learned how to deal with this problem.

Wanting to broaden our perspectives we should explain why it is "easy" to solve the first equations, second, third and fourth degree and why it is much less easy to solve those of higher than fourth degree.

The problem is not due to the simple fact that the analysis becomes more complicated, but that when the degree 4 is exceeded then you have to change your way of dealing with the problem, because the techniques of elevation and root extraction are no longer enough.

Mathematicians are more icastic and claim that the problem is no longer solvable in radicals. The problem becomes too technical, to be appreciated without an excessive effort.

We will therefore follow the thread that inspired the great Italian algebraists of the Renaissance period.

We have moved away from the Vedic conception of Mathematics and we would like to resume the thread of our thoughts starting from the way in which the algebraic equations were treated in the Vedic texts, thus returning to a less western view of the problem.

5.2   Vedic Conception and Second Degree Equation

We have already underscored that the Vedic doctrine of Mathematics departs from Western point of view in the sense that it is conceived as a "self consistent" discipline, sequentially developed from axioms and proceeding through theorems and lemmas. Rather, the problems are dealt with in parallel on the basis of assumptions, intuitions and prescriptions, which more than constituting a "corpus" are related to the specific of the problem under study. It is therefore extremely difficult to code a theoretical procedure and the methods to solve a class of problems translate in a kind of "tool kit". This procedure applies to the case of the second degree equations, whose solution is addressed in extremely interesting and quite different from the methods discussed so far.

We will proceed by illustrating some examples, the first of which is the following.

Example 14. *Let*

$$x + \frac{1}{x} = \frac{10}{3}. \tag{5.2.1}$$

The Vedic strategy is to exploit the symmetry of the equation, writing the known term in the same form of the term containing the unknown, things are easier to do than saying, we first note that

$$\frac{10}{3} = 3 + \frac{1}{3} \tag{5.2.2}$$

and then cast the original equation in the form

$$x + \frac{1}{x} = 3 + \frac{1}{3} \tag{5.2.3}$$

which suggests that the relevant solutions are $x = 3$ *and* $x = \frac{1}{3}$.

The examples reported below are useful to stress the flexibility of the method.

Example 15.

$$1) \qquad x + \frac{1}{x} = \frac{26}{5} = 5 + \frac{1}{5}. \qquad (5.2.4)$$

$$2) \qquad x + \frac{1}{x} = \frac{50}{7} = 7 + \frac{1}{7}. \qquad (5.2.5)$$

Example 16. *The equation*

$$(2x + 3) + \frac{1}{2x + 3} = \frac{50}{7} = 7 + \frac{1}{7} \qquad (5.2.6)$$

is just a variant of the first example which yields

$$2x + 3 = \begin{cases} 7 & \to & x = 2 \\ \dfrac{1}{7} & \to & x = -\dfrac{10}{7} \end{cases} . \qquad (5.2.7)$$

It is a slightly more elaborated case, which opens the way to further generalizations, as will be discussed later in this chapter.

We can reconcile western and Vedic conceptions by noting that any second degree equation $ax^2 + bx + c = 0$ can always be written as

$$x + \frac{1}{x} = x_+ + \frac{1}{x_+} = x_- + \frac{1}{x_-} \qquad (5.2.8)$$

where

$$x_\pm = \frac{-b \pm \sqrt{b^2 - 4ac}}{2a}, \qquad \frac{1}{x_\pm} = \frac{a}{c} x_\mp \qquad (5.2.9)$$

The "general" conclusion we may draw from the previous analysis is given below:

"The second degree equations can always be written as the equality of the sum of two terms reciprocal, the relative roots are such that the reciprocal of each corresponds to the conjugate of the other".

Although it may appear pretentious, we can say that the Vedic formulation contained, at least at the level of intuition, the general conception of complex numbers, we discussed in the last two chapters.

A different method, ascribed to the Vedic technicalities, is based on the already discussed factorization procedure that we will illustrate starting with the following example.

Example 17. *Let*

$$x^2 + 7x + 10 = x^2 + 2x + 5x + 10 = x(x+2) + 5(x+2) = (x+5)(x+2).$$
$$(5.2.10)$$

The roots of the equation $x^2 + 7x + 10$ are therefore given by -5 and -2.

The strategy underlying the factorization technique is that of enucleating, from the trinomial, a binomial to be put as common term.

The further examples reported below allow to become more familiar with this technique.

Example 18.

1) $2x^2 + 5x + 2 = 2x^2 + 4x + x + 2 = 2x(x+2) + (x+2) = (x+2)(2x+1).$
$$(5.2.11)$$

2) $4x^2 + 12x + 5 = 4x^2 + 2x + 10x + 5 = 2x(x+2) + 5(2x+1) = (2x+5)(2x+1).$
$$(5.2.12)$$

3) $12x^2 + 13x - 4 = 12x^2 + 16x - 3x - 4 = 3x(4x-1) + 4(4x-1) = (3x+4)(4x-1).$
$$(5.2.13)$$

5.3 Vedic Doctrines and Third Degree Equation

Even in the case of third degree equations the methods suggested by ancient Indian texts offer interesting ideas, although they should be seen in the perspective of the previous paragraph, that is, the absence of a unifying procedure. The extension of the method to the third degree equations is fairly straightforward, as illustrated below.

We consider the equation

$$x^2 + \frac{1}{x} = \frac{126}{5}, \tag{5.3.1}$$

by writing the second term as

$$\frac{126}{5} = 25 + \frac{1}{5} \tag{5.3.2}$$

we can rearrange the original equation as

$$x^2 + \frac{1}{x} = 25 + \frac{1}{5}, \tag{5.3.3}$$

thus finally arguing that the relevant solution is $x = 5$.

We can make further progress by including the remaining roots. To this aim we remind that in canonical form the previous equation writes

$$x^3 - \frac{126}{5}x + 1 = 0 \tag{5.3.4}$$

therefore, being the sum and the product of the roots of a generic equation $ax^3 + bx^2 + cx + d = 0$, given respectively by $-\frac{b}{a}$ and $-\frac{d}{a}$, we find

$$x_1 + x_2 + x_3 = 0, \qquad\qquad x_1 x_2 x_3 = -1 \tag{5.3.5}$$

so, by taking $x_3 = 5$, we get

$$x_1 + x_2 = -5, \qquad\qquad x_1 x_2 = -\frac{1}{5} \tag{5.3.6}$$

which yields the second degree equation, with solution

$$x_1 = \frac{-25 + \sqrt{645}}{10}, \qquad\qquad x_2 = \frac{-25 - \sqrt{645}}{10}. \tag{5.3.7}$$

The general conclusion we may draw from the previous discussion is
that any third degree algebraic equation expressible as

$$x^2 + \frac{1}{x} = a^2 + \frac{1}{a} \qquad (5.3.8)$$

admits the solutions

$$x_1 = a, \qquad x_2 = \frac{-a^2 + \sqrt{a^4 + 4a}}{2a}, \qquad x_2 = \frac{-a^2 - \sqrt{a^4 + 4a}}{2a}.$$
$$(5.3.9)$$

Although the previous method is not mentioned in the Vedic prescriptions, we have ascribed it in "ex officio" that simple generalization of the one relating to the solution of the second degree equations.

Let's go back to what has actually been handed down, considering
the equation

$$x^3 - 3x^2 + 4 = 0 \qquad (5.3.10)$$

and write it as

$$x^3 + x^2 - 4x^2 - 4x + 4x + 4 = x^2(x+1) - 4x(x+1) + 4(x+1) = 0 \quad (5.3.11)$$

it can be therefore straightforwardly factorized, thus ending up with

$$(x^2 - 4x + 4)(x + 1) = (x - 2)^2(x + 1). \qquad (5.3.12)$$

The same procedure can be applied to the equation

$$4x^3 + 7x + 4 = 0 \qquad (5.3.13)$$

easily factorized as

$$4x^3 + 2x^2 - 2x^2 - x + 8x + 4 = 2x^2(2x + 1) - x(2x + 1) + 4(2x + 1)$$
$$= (2x^2 - x + 4)(2x + 1) = 0.$$
$$(5.3.14)$$

Let us now discuss a more general procedure allowing to frame the
technique in more general terms. We consider therefore the third
degree trinomial $ax^3 + bx^2 + d$ and determine the conditions to cast

it in the product of, e.g., a second degree trinomial times a binomial. We rearrange the polynomial as

$$ax^3 + bx^2 + d = ax^3 + ex^2 + fx^2 + gx + hx + d$$
$$= x^2(ax + e) + x(fx + g) + hx + d. \tag{5.3.15}$$

Regarding this example, we argue that to end up with a factorization, the following conditions should be satisfied

$$a = f = h, \qquad e = g = d, \qquad e+f = b, \qquad g+h = 0. \tag{5.3.16}$$

In absence of the quadratic term, $ax^3 + cx + d$, the factorization proceeds as

$$ax^3 - ex^2 + ex^2 + gx + hx + d = x^2(ax - e) + x(ex + g)hx + d = 0 \tag{5.3.17}$$

thus getting for the factorization conditions

$$a = e = h, \qquad -e = g, \qquad g + h = c. \tag{5.3.18}$$

The procedure is not particularly laborious, but only applies in certain (fortunate) conditions, we will see in the following how the concepts discussed in this last section can be framed in a wider context.

5.4 A More General Definition of "Root"

The dream of any algebrist till the 19th century has been that of reducing the search of the roots of a n-th degree algebraic equation to the extraction of a root. The same as it happens when the search for the solutions of a second degree equation is reduced to the extraction of a square root.

The "trick" holds in general terms for equation up to the 4-th degree only but a tremendous amount of Mathematics was later developed to prove that algebraic equations with degree larger than 4 cannot be solved "in radicals"[3], which means by the use of the ordinary algebraic operations with the inclusion of the extraction of

---

[3]See J. Derbshire "Unknown Quantities" A Real and Imaginary History of Algebra, A Plume Book, Penguin Group, New York, 2007, Chapters 3 and 4.

roots. A revolutionary transformation, in the way of conceiving the algebraic equations, was later needed to state the conditions to be fulfilled to guarantee the solution by radicals of an arbitrary order algebraic equation.

The Ruffini-Abel Theorem (stating the impossibility of solving in radicals general polynomial of degree 5 and higher) and the Galois Theory (stating the conditions for the solubility in radicals) are milestones in the Mathematical thought. Unfortunately this book is not the proper location to deal with these topics, which requires a suitable discussion on the Group Theory.

We can however speculate on an extension of the concept (or better of the formalism) of root. We have already seen that the notation

$$x = {}^{(3,1_b)}\!\!\sqrt{-c} \tag{5.4.1}$$

represents the solution of the third degree equation

$$x^3 + bx + c = 0, \tag{5.4.2}$$

we can therefore also state that the following generalization of power exponent

$$y^{(3,1_b)} = y^3 + by. \tag{5.4.3}$$

The same holds for second degree polynomials, written as

$$y^{(2,1_b,0_c)} = y^2 + by + c \tag{5.4.4}$$

and can be extended to higher order polynomials, according to the obvious generalization

$$y^{(m,(m-1)_{b_1},\ldots,1_{b_{(m-1)}},0_{b_m})} = y^m + b_1 y^{(m-1)} + \cdots + b_{(m-1)} y + b_m. \tag{5.4.5}$$

We have started from what is called "ultra-radical" a concept sporadically appearing in the western mathematical literature[4] in regard to the solution of the fifth degree equation. We can accordingly introduce the concept of "ultra-power", denoting the inverse operation

---

[4]See e.g. Eric W. Weisstein, Ultraradical, MathWorld-A Wolfram Web Resource http://mathworld.wolfram.com/Ultraradical.html.

of ultra-radical.

Accordingly we can say that the solution of the ultra-power equation

$$y^{(5_b,3_c,0_d)} = 0 \tag{5.4.6}$$

is nothing but

$$y = {}^{(5_b,3_c)}\!\sqrt{-d} \tag{5.4.7}$$

Put in these terms, it seems that we are suggesting that being the inverse of an ultra-power is an ultra-radical, the problem of seeking the solution of an algebraic equation is solved in a fairly natural way. Obviously this is not the case, it is only a different way of formulating the question and the explicit computation of an ultra-radical is equivalent to the search for the root of an equation by ordinary means.

It is indeed evident that

$$^{(2,1_b)}\!\sqrt{-c} = \begin{cases} y_+ = \dfrac{-b + \sqrt{b^2 - 4c}}{2} \\ y_- = \dfrac{-b - \sqrt{b^2 - 4c}}{2}. \end{cases} \tag{5.4.8}$$

Hidden in the formalism, however, it is always possible to find some advantage.

Let's try, for example, to define the solution of the following fourth degree equation

$$(x^2 + dx)^2 + b(x^2 + dx) = c. \tag{5.4.9}$$

According to the just outlined formalism we can cast the previous equation in terms of ultra-power as

$$\left(x^{(2,1_d)}\right)^{(2,1_b)} = c. \tag{5.4.10}$$

The relevant ultra-radical solution reads

$$x = {}^{(2,1_b)}\!\sqrt{{}^{(2,1_d)}\!\sqrt{c}} \tag{5.4.11}$$

which is just a concise way of defining the solution of a bi-quadratic equation, namely

$$x = {}^{(2,1_b)}\!\!\sqrt{\frac{-d + \sqrt{d^2 + 4c}}{2}} = \frac{-b + \sqrt{b^2 + 2\left(-d + \sqrt{d^2 + 4c}\right)}}{2}.$$

$$(5.4.12)$$

The other solutions are defined by introducing the appropriate signs.

An alternative way of solving the above fourth equation is the procedure illustrated in Section 5.2, namely

$$y - \frac{1}{y} = y_\pm + \frac{1}{y_\pm}, \qquad y = \frac{x^2 + dx}{\sqrt{c}}, \qquad y_\pm = \frac{-\frac{b}{\sqrt{c}} \pm \sqrt{\frac{b^2}{c} + 4}}{2}.$$

$$(5.4.13)$$

The formalism of ultra-power can be further stretched, indeed by setting

$$x + \frac{1}{x} = x^{(1,-1)} \qquad\qquad (5.4.14)$$

and finding therefore

$$x + \frac{1}{x} = \alpha + \frac{1}{\alpha} \to x^{(1,-1)} = \alpha^{(1,-1)} \to x = {}^{(1,-1)}\!\!\sqrt{\alpha + \frac{1}{\alpha}} = \begin{cases} a \\ \dfrac{1}{\alpha} \end{cases}.$$

$$(5.4.15)$$

which also yields

$$x^2 + \frac{1}{x} = \alpha^2 + \frac{1}{\alpha} \to x^{(2,-1)} = \alpha^{(2,-1)}$$

$$\to x = {}^{(2,-1)}\!\!\sqrt{\alpha^2 + \frac{1}{\alpha}} = \begin{cases} \dfrac{-\alpha + \sqrt{\alpha^4 + 4\alpha}}{2\alpha} \\[2mm] \alpha \\[2mm] \dfrac{-\alpha - \sqrt{\alpha^4 + 4\alpha}}{2\alpha} \end{cases}.$$

$$(5.4.16)$$

These last identities, although very simple, highlight the flexibility of the concept of ultra-radical and ultra-power illustrated so far. We will return briefly to the use of this formalism by referring to solutions of the fifth degree equations, but for now we will be satisfied with a less "innovative" treatment.

5.5   Fifth Degree Polynomials

We completed the previous chapter with the derivation of a series of algorithms, related to the solution of algebraic equations of arbitrary degree, which, in essence, used a procedure aimed at reducing a given algebraic equation to another of lower order, with known solution. As underscored in the previous section, this procedure works up to the fourth degree.

We have also noted that this hope has guided for centuries the search for the (exact) solution formula of algebraic equations of arbitrary degree, we have also mentioned that this is not possible, at least when using radicals. We would like to convey that this does not mean that a general solution is not possible, but, in this case too, entirely new chapters of Mathematics had to be developed.

To get a very first idea of how the strategy of looking for the solution of an algebraic equations in radicals developed in the course of centuries, we use the so-called Viété method[5], for solving the equations of the second degree, which it is essentially the method of complement to the square, seen from a slightly different point of view.

The solution procedure consists in placing $x = y + z$ in the second degree equation at $ax^2 + bx + c = 0$, obtaining so

$$a(y^2 + z^2) + (2ay + b)z + by + c = 0. \qquad (5.5.1)$$

Our goal is that of obtaining a transformation which reduces the

---

[5]See e.g. F. Cajori, A History of Mathematical Notations, New York, Dover Publications, 1993 or F. Ritter, François Viéte, inventeur de l'algébre moderne, 1540-1603. Essai sur sa vie et son oeuvre, in "Revue occidentale philosophique, sociale et politique", 10, 1895. If one is interested to learn directly from the original books can look at Francoise Vieté, De aequationum recognitione et emendatione tractatus duo, published by Alexander Anderson, 1615.

problem of finding a solution to the extraction of a square root. We impose therefore that the coefficient multiplying the first degree in z term vanishes, thus finding

$$2ay + b = 0 \Rightarrow y = -\frac{b}{2a} \tag{5.5.2}$$

which, once replaced in the original equation, yields

$$az^2 + c - \frac{b^2}{4a} = 0 \tag{5.5.3}$$

thus ending up with the already known result.

The inspection of the just outlined Viété method allows at last two conclusions:

1) A second degree equation can be written in reduced form or without the first degree term through a suitable transformation of the variable;

2) The solution of a second degree equation is traced back to a first one in the unknown z and to subsequent extraction of a square root.

We will see below how the method just described is fundamental for the study of algebraic equations of third degree and beyond. We therefore consider the following third-degree algebraic equation, which we will say complete

$$ax^3 + bx^2 + cx + d = 0. \tag{5.5.4}$$

We could reach its solution, noting that this can be obtained by generalizing the operation of root extraction, and, using the ultra-radical formalism, we can write that

$$x = \left(^{3,2\frac{b}{a},1\frac{c}{a}}\right)\sqrt{-\frac{d}{a}}. \tag{5.5.5}$$

The previous writing would represent a solution of the problem, if we knew how to calculate an ultra-radical that is, if we had defined its properties rigorously. We will try to define a method for providing a calculation algorithm, using what we know about arbitrary algebraic equations. We carry out a Vieté type transformation and get the original third degree equation in the reduced form

$$R(t) = t^3 + pt + q = 0,$$

$$x = t - \frac{b}{3a}, \qquad p = -\frac{1}{3}\left(\frac{b}{a}\right)^2 + \frac{c}{a}, \qquad q = -2\left(\frac{b}{3a}\right)^3 - \frac{bc}{3a^2} + \frac{d}{a}.$$
$$(5.5.6)$$

In terms of ultra-radicals the relevant solution reads

$$t = {}^{(3,1}R\sqrt{-q}. \tag{5.5.7}$$

The use of the method of nested radicals yields

$$t = \sqrt[3]{-pt - q} \tag{5.5.8}$$

leading to the recursive procedure

$$t_{n+1} = \sqrt[3]{-pt_n - q}, \qquad\qquad t_0 = 0 \tag{5.5.9}$$

which yields an effective tool to evaluate ultra-radicals of order 3.

The following Fig. 5.3 shows how the use of the iterative procedure works to evaluate ultra-radical $^{(3,1}2.5}\sqrt{4}$. The convergence of the method appears rather slow. It is worth noting that $R(t_n)$ approaches values near zero for a number of iterations larger than 10.

So far practically nothing new has emerged, compared to what was discussed at the end of the previous chapter. A non-approximate procedure generalizing that adopted for the second degree equation is accordingly necessary.

In the spirit of Vieté we set $t = u + v$ and find

$$t^3 = 3uv(u+v) + (u^3+v^3) = 3uvt + (u^3+v^3) \Rightarrow t^3 - 3uvt - (u^3+v^3) = 0$$
$$(5.5.10)$$

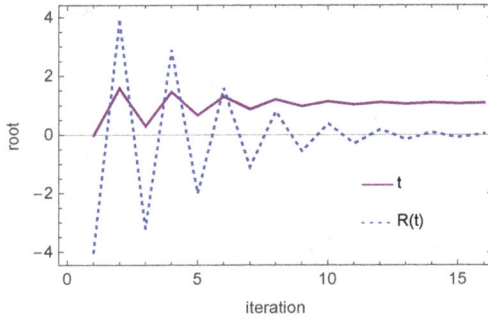

Figure 5.3: Use of the nested radicals procedure to evaluate the ultra-radical $^{(3,12.5)}\sqrt{4}$. Purple line: iteration for the positive root; Blue dashed line: $R(t)$ vs. the iteration number.

and use the transformations

$$3uv = -p, \qquad u^3 + v^3 = -q \qquad and \qquad u^3 = \zeta, \qquad v^3 = \eta \tag{5.5.11}$$

to get the following system of equations

$$\zeta\eta = -\frac{p^3}{27}, \qquad\qquad \zeta + \eta = -q \tag{5.5.12}$$

which yields the unknown

$$\zeta = \frac{-q - \sqrt{q^2 + 4\dfrac{p^3}{27}}}{2}, \qquad\qquad \eta = \frac{-q + \sqrt{q^2 + 4\dfrac{p^3}{27}}}{2} \tag{5.5.13}$$

as solutions of the second degree equation

$$\sigma^2 + q\sigma - \frac{p^3}{27} = 0. \tag{5.5.14}$$

According to this last identity the solution of the reduced cubic equation (or better one of its roots) can be expressed as

$$x = \sqrt[3]{\zeta} + \sqrt[3]{\eta}. \tag{5.5.15}$$

The roots of a third degree equation depends on that of a second degree equation. The solution formula that we have obtained turns out to be a sort of generalization of the Babylonian method of the square complement and we could define it, perhaps in a daring way, as a cube completion. The method is called Cardanico, in honor of the Italian mathematician Gerolamo Cardano who first described it in his book "Ars Magna" even if the real discoverer of the formula was an algebraist, always Italian, Niccoló Fontana called Tartaglia, who had a long correspondence with Cardano[6]. The big ideas are more and more fruitful of how it can be believed and go far beyond what is foreseen by the discoverer himself. In the following we will see how Tartaglia's idea can be further exploited.

We have so far seen that

$$^{(3,3}R)\sqrt{-q} = \sqrt[3]{\eta} + \sqrt[3]{\zeta} \qquad (5.5.16)$$

and since (η, ζ) are the solutions of a second degree equation, we could for example ask ourselves if there is one fifth degree equation that has a solution of the form

$$x = \sqrt[5]{x_+} + \sqrt[5]{x_-} \qquad (5.5.17)$$

where $x_\pm$ are solutions of the second degree equation $x^2 = \alpha x + \gamma$. To establish how this equation is made we proceed by raising both members of the previous expression to the fifth power and, using the fact that $x_+ x_- = \gamma$ and $x_+ + x_- = \alpha$, we get

$$x^5 - 5\sqrt[5]{\gamma}x^3 + 5\sqrt[5]{\gamma^2}x = \alpha. \qquad (5.5.18)$$

The explicit derivation requires some algebra, we invite the reader to try by his/her own. If the derivation takes too long, we have added the following hint. We note that the explicit derivation is obtained after noting that

$$x^5 = x_+ + 5x_+^{\frac{4}{5}}x_-^{\frac{1}{5}} + 10x_+^{\frac{3}{5}}x_-^{\frac{2}{5}} + 10x_+^{\frac{2}{5}}x_-^{\frac{3}{5}} + 5x_+^{\frac{1}{5}}x_-^{\frac{4}{5}} + x_- . \qquad (5.5.19)$$

---

[6]See J. Derbshire "Unknown Quantities" A real and Imaginary History of Algebra, A Plume Book, Penguin Group, New York, 2007, chapters 3 and 4.

The rhs can be rearranged as

$$x^5 = x_+ + x_- + 5\sqrt[5]{x_+ x_-}\left(x_+^{\frac{3}{5}} + 2x_+^{\frac{2}{5}}x_-^{\frac{1}{5}} + 2x_+^{\frac{1}{5}}x_-^{\frac{2}{5}} + x_-^{\frac{3}{5}}\right)$$

$$= x_+ + x_- + 5\sqrt[5]{x_+ x_-}\left[\left(x_+^{\frac{1}{5}} + x_-^{\frac{1}{5}}\right)^3 - \sqrt[5]{x_+ x_-}\left(x_+^{\frac{1}{5}} + x_-^{\frac{1}{5}}\right)\right].$$

$$(5.5.20)$$

According to previous relations we can also conclude that the following identity in terms of ultra-radicals holds

$$^{(4,5_A,5_B)}\!\!\sqrt{\alpha} = \sqrt[5]{x_+} + \sqrt[5]{x_-}, \qquad A = -5\sqrt[5]{\gamma}, \qquad B = 5\sqrt[5]{\gamma^2}.$$

$$(5.5.21)$$

Well, the equation just obtained is a fifth degree equation whose solution depends on the roots of a second degree. The relative solution can therefore be obtained by using successive extraction operations of roots (square and quintuple) and therefore without introducing concepts going beyond the operation of radicals. Unfortunately not all quintic (or fifth-degree equations) can be reduced to the previous form, which remains a particular case (said De Moivre type equations) of a "much" more complicated problem. We invite the reader to prove that the seventh and eighth degree De Moivre equations are

$$x^7 - 7cx^5 + 14c^2x^3 - 7c^3x - \alpha = 0,$$
$$x^8 - 8cx^6 + 20c^2x^4 - 16c^3x^2 + 2c^4x - \alpha = 0, \qquad (5.5.22)$$
$$c = \sqrt[n]{\gamma}, \quad n = 7, 8$$

and that the coefficients of a general De Moivre equation can be specified through the following De Moivre triangle (see Table 5.2).

In the rest of the Chapter we complete the discussion developed so far, by expanding the examples we have just analyzed.

5.6   Algebraic Equations, Trigonometry ... ... and More

We have already highlighted the link between second degree algebraic equations, geometry and trigonometry, which we will now

Table 5.2: **De Moivre triangle.**

1							
1							
1	-2						
1	-3						
1	-4	2					
1	-5	5					
1	-6	9	-2				
1	-7	14	-7				
1	-8	20	-16	2			
1	-9	27	-30	9			
1	-10	35	-50	25	-2		
1	-11	44	-77	55	-11		
1	-12	54	-112	105	-36	2	
1	-13	65	-156	182	-91	13	
1	-14	77	-210	294	-196	49	-2

try to deepen. Although not explicitly stated the considerations we have made the solutions of a second degree equation are given to us to conclude that its roots can be elaborate using trigonometric formulas; in fact suppose that $\Delta < 0$, and let us say

$$x_+ = Ae^{i\phi}, \qquad\qquad x_- = Ae^{-i\phi}. \qquad (5.6.1)$$

We can determine A and ϕ by taking into account that

$$x_+ x_- = \frac{c}{a} \Rightarrow A = \sqrt{\frac{c}{a}} \qquad (5.6.2)$$

and that

$$\frac{x_+ + x_-}{2} = A\cos\phi \;\Rightarrow\; A\cos\phi = -\frac{b}{2a},$$

$$\frac{x_+ + x_-}{2i} = A\sin\phi \;\Rightarrow\; A\sin\phi = \frac{\sqrt{|\Delta|}}{2a} \;\Rightarrow\; \tan\phi = -\frac{\sqrt{|\Delta|}}{b}.$$

$$(5.6.3)$$

These solutions can be represented on the Argand-Gauss plane as shown in the following figure. Obviously the same procedure applies to the case of the cubic equation solved in "cardanic" form and in fact we can write

$$x = \sqrt[3]{x_+} + \sqrt[3]{x_-} = 2\sqrt[3]{A}\cos\frac{\phi}{3} \qquad (5.6.4)$$

which is valid if $q^2 + 4\frac{p^3}{27} < 0$. If we furthermore keep into account that

$$\sqrt[3]{1} = \begin{cases} 1 & = & 1 \\[2mm] \dfrac{-1 + i\sqrt{3}}{2} & = & e^{\frac{2i\pi}{3}}, \\[2mm] \dfrac{-1 - i\sqrt{3}}{2} & = & e^{-\frac{2i\pi}{3}}, \end{cases} \qquad (5.6.5)$$

we can get the trigonometric form of a cubic equation as

$$x_2 = 2\sqrt[3]{A}\cos\frac{\phi + 2\pi}{3}, \qquad\qquad x_3 = 2\sqrt[3]{A}\cos\frac{\phi - 2\pi}{3}. \quad (5.6.6)$$

All equations in the form of De Moivre admit similar solutions. Given therefore an equation of this kind we can represent its solutions on the border of a circle.

Before closing this section, we will return to the Vedic treatment of the problem, trying to put together the considerations of the previous paragraph with those developed in this.

We consider the equation

$$x^3 + 3x = 4, \qquad (5.6.7)$$

we easily establish that one of its solutions is 1 and that the others are complex. From the Cardanic solution formula we also infer that

$$x = \sqrt[3]{2 + \sqrt{5}} + \sqrt[3]{2 - \sqrt{5}} \tag{5.6.8}$$

and furthermore that[7]

$$\sqrt[3]{2 + \sqrt{5}} + \sqrt[3]{2 - \sqrt{5}} = 1. \tag{5.6.9}$$

An identity which is hard to establish independently of the properties of a third degree algebraic equation. Should anyone who wish to attach an esoteric meaning to the previous result we could also observe that the previous one combination of radicals belong to the family of "mystical radicals", as they are related to the golden ratio through identity

$$\sqrt[3]{2 \pm \sqrt{5}} = \left\{ \begin{array}{l} \phi \\ -\phi' \end{array} \right. \tag{5.6.10}$$

which are not hard to state and we invite the reader to prove both.

A further pastime we propose is to try the following identities and decide which is incorrect, without any recourse to the pocket computer,

$$\sqrt[9]{38 \pm 17\sqrt{5}} = \sqrt[6]{9 \pm 4\sqrt{5}} = \sqrt[3]{2 \pm \sqrt{5}} \tag{5.6.11}$$

and

$$\sqrt[3]{7 \pm 5\sqrt{2}} = 1 \pm \sqrt{2} \tag{5.6.12}$$

which is linked to the silver ratio and it is not hard to prove.

5.7 Fourth Degree Polynomials

In the previous paragraph we learned that the solution of a third degree equation is "reducible", through the Tartaglia-Cardano-De Moivre (TCD) method, to a second degree. In addition we have also

---

[7]See e.g. Thomas J. Osler, Cardan polynomials and the reduction of radicals, Mathematics Magazine.

seen that an entire class of equations (above the third) can be reduced to cardanic forms. We now hope to get there something similar, at least from a methodological point of view, with regard to the fourth equations degree. So, as we already know, the transformation of Vieté allows us to eliminate the third degree term, in the following we consider the "canonical" form

$$x^4 + px^2 + qx + d = 0, \qquad (5.7.1)$$

set

$$x = u + v + z \qquad (5.7.2)$$

and eventually note that

$$
\begin{aligned}
x^2 &= u^2 + v^2 + z^2 + 2(uv + vz + zu) \\
\Rightarrow x^2 - (u^2 + v^2 + z^2) &= 2(uv + vz + zu) \\
\Rightarrow [x^2 - (u^2 + v^2 + z^2)]^2 &= [2(uv + vz + zu)]^2 \qquad (5.7.3) \\
x^4 - 2(u^2 + v^2 + z^2)x^2 + (u^2 + v^2 + z^2)^2 & \\
&= 4(u^2v^2 + v^2z^2 + z^2u^2) + 8uvz(u+v+z).
\end{aligned}
$$

Albeit not evident, the problem has been solved, to this aim we write the last line of the previous equation as

$$x^4 - 2(u^2 + v^2 + z^2)x^2 - 8uvzx + (u^2 + v^2 + z^2)^2 - 4(u^2v^2 + v^2z^2 + z^2u^2) = 0. \qquad (5.7.4)$$

Finally, by setting

$$
\begin{aligned}
u^2 + v^2 + z^2 &= -\frac{p}{2}, & uvz &= -\frac{q}{8}, \\
(u^2 + v^2 + z^2)^2 - 4(u^2v^2 + v^2z^2 + z^2u^2) &= r, & & (5.7.5) \\
(u^2v^2 + v^2z^2 + z^2u^2) &= \left(\frac{p^2 - 4r}{16}\right) &
\end{aligned}
$$

and

$$U = u^2, \qquad V = v^2, \qquad Z = z^2, \qquad (5.7.6)$$

we end up with the system

$$\begin{cases} U + V + Z = -\dfrac{p}{2}, \\[2mm] UV + VZ + ZU = \dfrac{p^2 - 4r}{16}, \\[2mm] UVZ = \dfrac{q^2}{64} \end{cases} \qquad (5.7.7)$$

which eventually leads to the third degree equation

$$y^3 - \frac{p}{2}y^2 + \frac{p^2 - 4r}{16}y - \frac{q^2}{64} = 0. \qquad (5.7.8)$$

The problem has therefore been solved in very general terms. The explicit writing of the relative solutions is very involved and would add little to what we wanted to communicate.

We believe it appropriate to close this section by trying to summarize the solution method[8] proposed by Ludovico Ferrari[9] (a pupil of Cardano), who arrives at the solution, using four steps.

For simplicity, let's consider a specific example, that is, the search for solutions to the equation

$$x^4 + 4x^2 + 36 = 60x. \qquad (5.7.9)$$

The problem will be addressed with a technique that recalls the square complement:

1) We operate in such a way as to transform the first member into a perfect square, in the present case by adding and subtracting $8x^2$ we find

$$x^4 + 4x^2 + 8x^2 + 36 = 8x^2 + 60x \;\Rightarrow\; (x^2 + 6)^2 = 8x^2 + 60x. \quad (5.7.10)$$

---

[8]See   http://www.matmedia.it/Antologia/I%20grandi%20momenti/Formula%20di%20Ferrari%20per%20le%20quartiche/la_formula_di_ferrari_per_le_qua.htm.

[9]It is curious to quote that, in his opera "Ars Magna", Gerolamo Cardano makes reference to the Ferrari Method for the solution of the fourth degree equation. Just quoting his words: *"The solution of the fourth degree equation was obtained by Ludovico Ferrari on my explicit request"*. Not fair, isn't it? Ludovico Ferrari at the time of his discovery was seventeen and died poisoned.

2) An auxiliary variable y is introduced so as to leave the first member in the form of a square, namely, by using eq. (5.7.10),

$$(x^2+6+y)^2 = (x^2+6)^2+2y(x^2+6)+y^2 = 8x^2+60x+2y(x^2+6)+y^2$$
$$= 2(4+y)\left(x^2+\frac{30}{4+y}x+\frac{y(y+12)}{2(4+y)}\right);$$

$$(5.7.11)$$

3) The term in square brackets on the rhs of (5.7.11) is a second degree polynomial in x. It reduces to a perfect square if the associated discriminant is zero, namely

$$\Delta(y) = \left(\frac{30}{4+y}\right)^2 - 2\frac{y(y+12)}{(4+y)}.\qquad(5.7.12)$$

The condition (5.7.12) fix a specific value of y, denoted by y^*. We can therefore write eq. (5.7.11) as

$$(x^2+6+y^*)^2 = 2(4+y^*)\left(x+\frac{1}{2}\frac{30}{4+y^*}\right)^2,\qquad(5.7.13)$$

which is just a biquadratic equation in the unknown x and eventually

$$x^2+6+y^* = \pm\sqrt{2(4+y^*)}\left(x+\frac{15}{4+y^*}\right).\qquad(5.7.14)$$

The comment to the procedure, we have outlined in steps $(1-3)$ can be summarized as: *the solution of a fourth degree equation depends on those of a third equation to find y^* and that of a bi-quadratic equation to obtain the solutions associated with the unknown x.*

The last example of this section is the solution of the quintic equation

$$x^5+x+a = 0\qquad(5.7.15)$$

which in terms of ultra-radicals reads

$$x = \sqrt[(5,1)]{-a}.\qquad(5.7.16)$$

This form of the solution (although with a different "symbolism") is called **Bring-Jerrard**[10], the calculation explicit of the ultra-radical is performed by placing

$$f(x) = x^{(5,1_1)} = x^5 + x \qquad (5.7.17)$$

and calculating the solution of our problem through the series development of $f^{-1}(-a)$ (i.e. as the inverse of the super-power of order 5), in this case the expansion writes

$$^{(5,1_1)}\!\!\sqrt{-a} = -\sum_{k=0}^{\infty} \binom{5k}{k} \frac{(-1)^k a^{4k+1}}{4k+1} = -a + a^5 - 5a^9 + 35a^{13} + \ldots$$

$$(5.7.18)$$

We have just mentioned this result, which is by no means straightforward, since it involves a further step in the evolution of the theory of algebraic equations[11].

The method consists in considering the unknown x of the quintic equation as an analytic function of the known term a and then proceeding by finding an analytical expansion of x in terms of a. We will not further comment on this aspect of the theory, which needs further notions out of the scope of this book.

We mixed old and modern and learned a number of things. Evidently the complete solutions of the high degree equations (which include the third and fourth) involve a considerable amount calculation effort. Even the general solution, if written so as to cover all possible cases, it becomes very little transparent. Yet Renaissance mathematicians had developed methods that allowed the calculation of its roots with remarkable precision and what made it the most

---

[10]The history is rather long and complex there have been many contributors and each one would deserve credit. For a fairly complete account see "The Legacy of Niels Henrick Habel", The Abel bicentennial 2002, Ed. by Olaf Arnfinn and Ragni Piene, Springer Verlag Berlin Heidelberg (2004). We also like to quote the contribution by an illustrious Italian Mathematician Giuseppe Belardinelli, Fonctions hypergeomètriques de plusieurs variables et rèsolution analytique des equations algèbriques gènèrales, Memorial des Sciences Mathèmatiques CXLV, Gauthier-Villars, Paris, 1960.

[11]See previous footnote.

exceptional was the notation that certainly didn't help. In the next paragraph we will mention these issues, trying to understand how difficult the art of calculation was and why it was important to find increasingly accurate methods.

In the next paragraph we will see to what extent.

5.8   Solutions and Approximations

The notation used by the Italian algebraists was not in fact dissimilar from those introduced by Arabic mathematicians, who classified 14 types of cubic equations[12] using verbal expressions; for example "Cubes and censuses and things equal to number" indicated the equation $x^3 + ax^2 + bx = c$.

Yet the methods used allowed to reach considerable approximations, corrected to the 10th decimal place. As an example, we report the solution of the equation "Cubes equal to things and numbers" ($x^3 = 3x + 1$) solved by Persian mathematician Biruni (actually Al-Birooni) around the year 1000. Biruni does not explain the technique used and provides the value of the unknown in sexagesimal form or $x = 1^{\circ}52'45''47'''13^{iv} = 1.879385262345679$. To verify its accuracy, we used the iterative method of *nested radicals methods* (NRM)

$$x_{n+1} = \sqrt[3]{1 + 3x_n}, \qquad\qquad x_0 = k \qquad (5.8.1)$$

and we have obtained the results reported in the following Figure 5.4. After 15 iterations we find a result very close to that of Biruni, after 30 we end up with a stable solution to 16-th decimal place.

Biruni also finds the solution 0.3472963734567901 of the equation $x^3 = 3x - 1$. We used the iterative method to obtain this value starting from numbers very close (but not identical) to the value given by

---

[12]Berggren, J. Lennart (1986). *Episodes in the Mathematics of Medieval Islam.* New York: Springer-Verlag. ISBN 0-387-96318-9.
See also M. Burgess Connor, A historical survey of methods of solving cubic equations, https://scholarship.richmond.edu/cgi/viewcontent.cgi?article=1113&context=masters-theses.

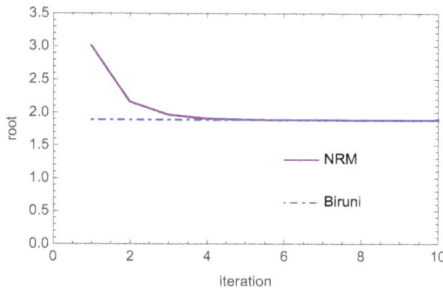

Figure 5.4: Comparison between Biruni's first equation solution and NRM approximation method by starting from $x_0 = 3$.

the Persian mathematician. The following Figure 5.5 shows the results of the iteration that are not "stable" and after a certain number of "tests" move towards the two other cubic solutions which, from the point of view of our algorithm, are very stable.

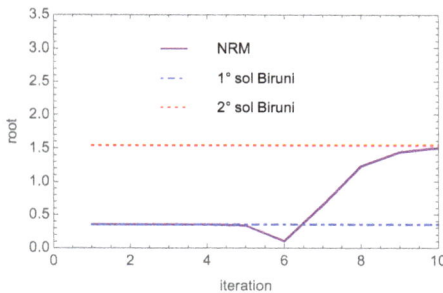

Figure 5.5: Comparison between Biruni's second equation solution and NRM approximation method by starting from $x_0 = 0.347293$.

Evidently, Arab mathematicians had approximation techniques of common use, not reported in the Biruni manuscript, which uses the aforementioned equations to solve the problem of building regular polygons of 9 and 18 sides.

These methods were imported in Italy by Fibonacci which provides the solution of the equation $x^3 + 2x^2 + 10x = 20$ with an approximation of ten decimal places[13]. In this case the nested radicals algorithm does not provide a reliable result, using instead a more reliable method (Newton's method) we obtain what is shown in Fig. 5.6, which reports the solution obtained by successive iterations. An approximate solution to the fourteenth decimal place is $x = 1.3688081078(21373)$, where the digits in parentheses were obtained on the basis of a numerical computation) is obtained with a low number of iterations ($n = 5$ is sufficient).

The comparison with the result provided by Fibonacci on a sexagesimal basis is given below

$$x = 1^\circ 22' 7'' 42''' 33^{iv} 4^v 40^{vi} = 1 + \frac{22}{60} + \frac{7}{60^2} + \frac{42}{60^3} + \frac{33}{60^4} + \cdots$$

$$= 1.3688081078(532)$$

$$(5.8.2)$$

which is remarkable, taking into account the computational tools of the time, and all the other difficulties associated with the notation itself.

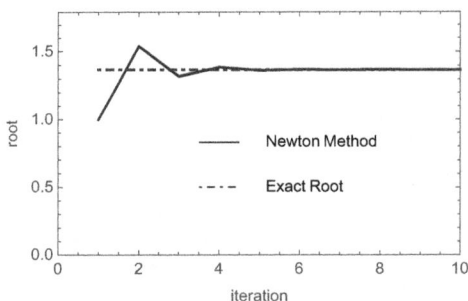

Figure 5.6: Fibonacci equation solution through the Newton method with $x_0 = 1$.

---

[13]To tell the truth the number of digits was 11, the last was wrong!!! Fair persons do not underscore this point, but we are not fair because we felt so much frustrated being barely able to find by "hand computing" the third digit (but in the end we cannot guarantee that the mistake was just due to our numerical computation).

Regarding the method, we can presume that the algorithmic procedure has been similar to the Archimedean method of the intersection of the conics and presumably it required a significant amount of arithmetic. We therefore tried ad apply this method to the solution of the Fibonacci equation, introducing the following auxiliary curve

$$y = \frac{1}{2p}x^2, \qquad\qquad pyx + 2py + 5x = 10 \qquad\qquad (5.8.3)$$

which, once inserted into the third degree equation, yields the curve

$$pyx + 2py + 5x = 10 \;\Rightarrow\; y = -\frac{5(x-2)}{p(x+2)} \qquad\qquad (5.8.4)$$

and the intersection between the two curves is the solution of the original equation, as reported in Fig. 5.7.

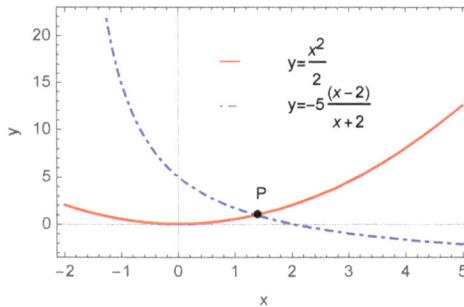

Figure 5.7: Fibonacci cubic equation solution by curves intersection method.

This technique is essentially the same that led Menecno to solve the problem of the duplication of the volume of the cube. The same procedure applied to Biruni's equations allows to solve the problem just as easily. The "tested" methods of solution in the Mediterranean area around 1000 years were probably these and they had their origin in the geometric concept of algebra developed by the Greeks. The geometric point of view continues also in our days and although the cubic problem has been examined in all its aspects, sometimes some valuable results are still produced.

Before concluding this chapter, we will mention the "geometry" of the cubic equation using what is shown in[14] Fig. 5.8, where we have reported the function $y = ax^3 + bx^2 + cx + d$.

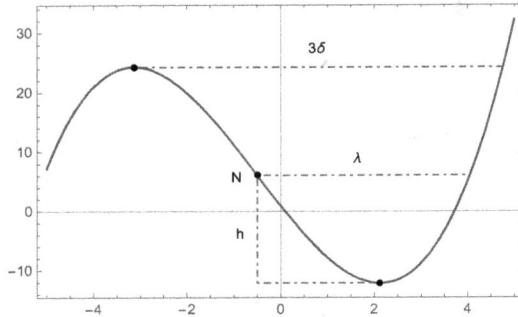

Figure 5.8: Cubic geometry.

The symmetry point of the curve is given by

$$N \equiv (x_N, y_N), \qquad\qquad x_N = -\frac{b}{3a} \qquad (5.8.5)$$

and provides the geometrical meaning of the Vieté transformation.

Further important parameters are reported below

$$\lambda^2 = 3\delta^2, \qquad\qquad h = 2a\delta^2, \qquad\qquad \delta^2 = \frac{b^2 - 3ac}{9a^2}. \qquad (5.8.6)$$

It also evident that

$$
\begin{aligned}
y_n^2 &> h^2, &\quad &\text{only one root } \in \mathbb{R} \\
y_n^2 &= h^2, &\quad &\text{three roots } \in \mathbb{R} \text{ (two coincident roots).} \quad (5.8.7) \\
y_n^2 &< h^2, &\quad &\text{three roots } \in \mathbb{R}
\end{aligned}
$$

Now let's consider the last case (three real roots) and try to see what is the geometric meaning of the trigonometric solutions, which

[14]R. W. D. Nickalls, The Mathematical Gazette n. 77, 1993, pp. 354–359.

are shown graphically in the Fig. 5.9, and can be expressed as

$$\alpha = x_N + 2\delta \cos\theta, \qquad\qquad \beta = x_N + 2\delta \cos\left(\theta + \frac{2}{3}\pi\right),$$

$$\gamma = x_N + 2\delta \cos\left(\theta + \frac{4}{3}\pi\right), \qquad\qquad \cos(3\theta) = -\frac{y_N}{h}.$$

$$(5.8.8)$$

We apply the geometrical method to the solution of the equation

$$x^3 - 7x^2 + 14x - 8 = 0. \qquad\qquad (5.8.9)$$

The use of the previous prescriptions yields

$$x_N = \frac{7}{3} \Rightarrow y_N \simeq -0.7407, \qquad\qquad \delta^2 = \frac{7}{9},$$

$$(5.8.10)$$

$$h \simeq 1.3718, \qquad\qquad \cos(3\theta) \simeq 0.5399 \Rightarrow \theta \simeq 19.1066,$$

which allows the derivation of the roots

$$\alpha = \frac{7}{3} + 2\sqrt{\frac{7}{9}} \cos(19.1066^\circ) = 4,$$

$$\beta = \frac{7}{3} + 2\sqrt{\frac{7}{9}} \cos(19.1066^\circ + \frac{2}{3}\pi) = 1, \qquad (5.8.11)$$

$$\gamma = \frac{7}{3} + 2\sqrt{\frac{7}{9}} \cos(19.1066^\circ + \frac{4}{3}\pi) = 2.$$

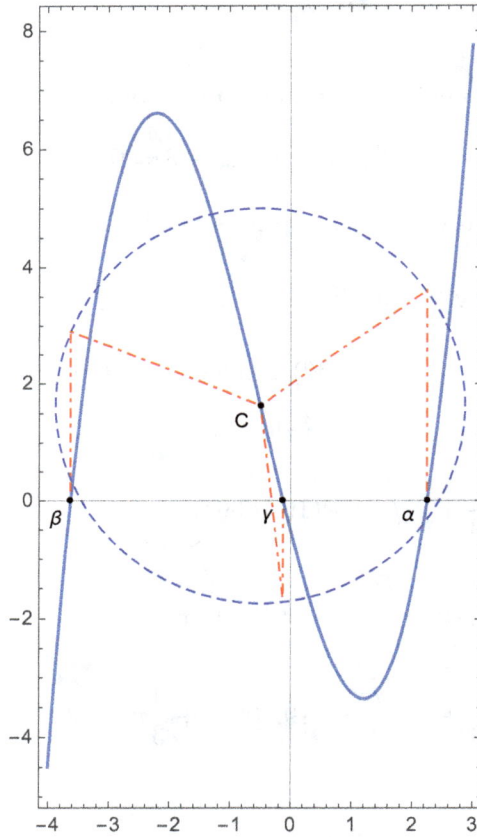

Figure 5.9: Cubic trigonometry.

The fourth degree equation did not find great applications in the ancient world, indeed, given the practical conception of the mathematics of those times, the search for its solutions was considered a waste of time, since the powers of 4 (and higher) seemed to have no geometric meaning.

The interest in third degree equations, on the other hand, has always been alive, and we report a sort of ruler designed for the solution of cubic equations.

The "device" works as follows: the curve is shown on a ruler graduated on a linear scale equation $y = x^3$ which is intersected with the straight line $y = -px - q$, the intersections obviously provide the solutions of the equation $x^3 = -px + -q$. The coefficients p and q are simply obtained from the intersection with the parallel axes of the ruler (see also Fig. 5.10).

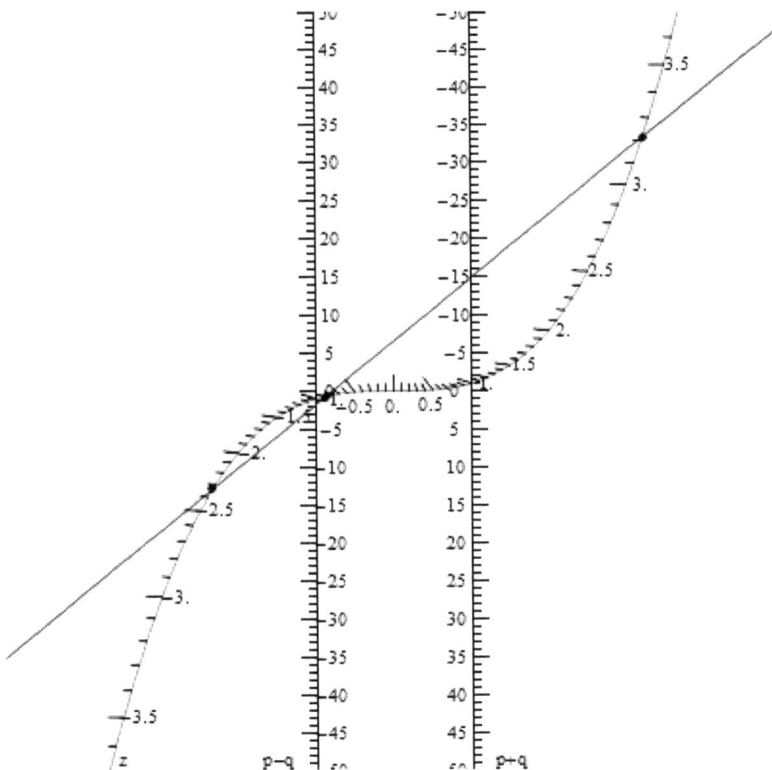

Figure 5.10: Ruler for cubic equation.

Chapter 6

The Two Souls of Mathematics and the Last of Vedas

6.1 Introduction

This chapter, although devoted to elementary aspects of Mathematics, requires some extra effort and demands for some abilities which bring us beyond (but not too much) the minimal formalism we have exploited so far. We will try to alleviate the computational pain, by offering our support whenever necessary.

Although not explicitly stated, what has been discussed in the previous chapters was, among other things, aimed at suggesting the idea of how Mathematics is a discipline very hard to be defined, or, better to say, very different from what is usually understood. Contrary to the widespread belief, popularized by the commonplace "Mathematics is not an opinion", it cannot be considered free from experimentation and the idea that it rests on eternal and immutable grounds it is, to a large extent, an illusion. One of the great intellectual revolutions (may be the greatest) of the last century was the end of hope that in mathematics, the criterion of non-contradiction is valid, and therefore demonstrable. The K. Gödel's work brought the end of this belief.

By expressing ourselves in span and reducing the matter in really gross terms, we could assert that there is no certainty that, within a system of axioms, one cannot prove two theorems, equally "true", which are the negation of the other[1]. This intellectual "scandal" closed Hilbert's program and posed questions of a philosophical nature, which then would have reverberated on other sciences, such as physics and on the claim of the formulation of the so-called "Theory of the Everything"[2].

Without further indulging in these matters, we could assert that although the illusion has been lost of the existence of the principle of non-contradiction, it is possible to continue doing Mathematics in the absence of such certainty. What history has shown is that "the principles of mathematical deduction and not only theories mathematics have undergone changes over the centuries"[3]. Furthermore, as an alternative to the "Greek" concept which during the last century has found its raison d'etre in the realization of Hilbert's program and in his followers of the Bourbaki current, there is an alternative way to do Math

"Equally efficient, with non-conflicting objectives and with methods capable of providing mutual help".
(P. Cartier)

---

[1] In less naïve terms, *the (second) Gödel Incompleteness Theorem* should be worded a sit follows: *"Be S a formal system which a) Contains the language of Arithmetics, b) Includes the Peano's axioms, c) Is coherent.* Regarding c), the coherence cannot be proved in *S.*

[2] The problem is still hot; Freeman Dyson (New York Review of Books, May 13, 2004) underscored that: *"Gödel's Theorem implies that pure mathematics is inexhaustible. No matter how many problems we solve, there will always be other problems that cannot be solved within the existing rules. ...because of Gödel's Theorem, Physics is inexhaustible too. The laws of physics are a finite set of rules, and include the rules for doing Mathematics, so that Gödel's Theorem applies to them"*.

[3] See P. Cartier, Mathemagics, Seminaire Lothangerien de Combinatoire n. 44, 2000, Article B44d.

Such complementary conceptions of mathematical thought have historical roots which deepened very far in the history of mankind culture. Despite the limited nature of topics covered during our discussion of the various forms of mathematics that unfolded over the centuries, we could appreciate the existence of two currents, roughly ascribable to Greek and "Eastern" (Assyro-Babylonian, Indian...). The latter perceived as if it were the antagonist of the western counterpart, which became an interpreter of the Greek tradition. There is no doubt that the western path to mathematics led, after the Renaissance, to a total detachment from hybrid concepts increasingly marking the territory of relevance towards a rigorist identity, which culminated in the already quoted Hilbert program. Nonetheless, gaps opened which during the 18th and 19th centuries had different interpreters, with the inclusion of venerable fathers like Euler, who used reckless operations on divergent series, Boole, who can be considered the founder of the symbolic calculation, Heaviside, who in turn created a symbolic calculation for electromagnetism, who lived in constant conflict with his contemporaries of a more orthodox orientation. Between the first and second half of the 19th century, a school of operationalists flourished (Glaisher, Crofton ... [4]) who managed to bring the symbols and operations between them to their extreme consequences and created the methodology that culminated in modern conceptions of umbral calculus. To a large extent this way of conceiving Mathematics influenced Physics. Many great physicists adopted methods tracing back to operational calculus. Dirac defined the Heaviside methods magic[5] and Feynman used (invented and reinvented) many aspects of this calculus for various issues in quantum electrodynamics[6].

---

[4]See the Essay P. Cartier, Mathemagics (A Tribute to L. Euler and R. Feynman), Seminaire Lotharingien de Combinatoire 44, 2000, Article B44d and Also J. Mikusiński, Operational calculus, 1, PWN & Pergamon 1987, (Translated from Polish); B. van der Pol, H. Bremmer, Operational calculus based on the two-sided Laplace integral, Cambridge Univ. Press, 1959.

[5]See H. S. Kragh, Dirac: A scientific biography, 2008 and D. H. Moore, Heaviside operational calculus: An elementary foundation (Modern analytic and computational methods in science and mathematics), Elsevier, 1971.

[6]R. P. Feynman, An operator calculus having applications in quantum electrodynamics, Phys. Rev., 84, 1951, pp. 108–128.

In this concluding chapter we will try to gather the threads of the discourse that has unfolded in the previous ones, returning to some points that, perhaps, have not attracted due attention. We will argue something more. We would also like to provide an idea of how the two "alternative" mathematics tendencies merged into the work of a mathematician who certainly represents the last great exponent of the Indian mathematical tradition and who yet it has made one of the major contributions to Mathematics of the last century. In fact, we will talk about Srinivasa Ramanujan whose work and specific skills are not easily confronted with those of the (great) mathematicians of the "western" tradition. We are not referring to the importance of the contributions, nor to technical skills, but to the completely original way, apparently devoid of any consequentiality logic, with which his research was conceived and presented. There are many legends about Ramanujan, which albeit to be taken with extreme cautions, reveal what has been the legend which his mathematical abilities arouse around his figure. It is said that once, when Ramanujan was a boy, somebody asked him the solution of the system

$$\sqrt{x} + y = 7, \qquad\qquad \sqrt{y} + x = 11, \qquad (6.1.1)$$

an "educated" person would reduce the problem to a fourth degree equation for y, namely

$$a(y)^2 - 22a(y) + 121 - y = 0, \qquad\qquad a(y) = y^2 - 14y + 49 \quad (6.1.2)$$

and to something analogous for x. If we follow this way, we have shown that the problem admits a solution, which requires finding the roots of a fourth degree equation. The answer by the young Ramanujan was

$$x = 9, \qquad\qquad y = 4. \qquad (6.1.3)$$

The example we have just discussed is not unlike those regarding the solution in vedic texts of second degree equations based on an intuition linked to the symmetry properties of the equation itself. The solution so obtained is incontrovertibly true, so naturally true that one can ask how come that I hadn't thought of that?

The number 1729 says nothing to almost all human beings, but for Ramanujan it was unique, since it represents[7] "The smallest number expressed as the sum of two cubes in two different ways"

$$1 + 12^3 = 9^3 + 10^3. \tag{6.1.4}$$

In this case it is difficult to imagine the mental processes that led to this conclusion. We are led to believe that the founding elements of the extreme originality of Ramanujan's thought are essentially two. Except for the immense natural talent, which put him on the same level of the greatest mathematicians of any time, Euler and Gauss, his cultural independence was born from a practically non-existent formal education and from being trained as an accountant.

At the time of Ramanujan's youth, in India, this profession was based on the calculation methods of the ancient Vedas. These methods, which he probably had refined, his natural inclination towards a mathematical thought made of intuition and ability to see the forms, rather than the formulas, brought to achievements so profound and rich that the consequences are still matter of active research. We will only touch on "elementary" and nonetheless particularly enlightening questions that allow us to understand how these are in perfect resonance with the Indian mathematical conception that we discussed in course of the first chapters.

We like to close this section with a question asked by Ramanujan himself:

"What is the value of the following expression?"

$$\sqrt{1 + 2\sqrt{1 + 3\sqrt{1 + 4\sqrt{1 + \ldots}}}} \tag{6.1.5}$$

---

[7]In view of eq. (6.1.4), can you explain why the cube root of 1729 is not an integer?

We will provide the solution later in this chapter and it will be so natural that it will be difficult to avoid asking yourselves:

"How did it come that I did not think before?"

6.2 Problems with Birthdays and Addresses

A few years ago, Italian television had a column in which a nobleman gave "bon ton" lessons. There is no doubt that from these lessons the public benefited enormously. During a memorable broadcast there was discussion about how to arrange guests at the table if there were ambassadors and an Apostolic Nuncio (a high ranked priest acting as Ambassador for the Vatican State) among the guests. After a careful examination the viewers have been informed about the correct definition of the positioning geometry, which it would have had to take into account the various relationships between the countries of origin of the ambassadors with the nation of the guest, while for the nuncio, by virtue of his ecumenical role, a privileged allocation should have been considered, free from geopolitical considerations.

A rude young man placed the nobleman in serious difficulty by asking how they should have changed provisions if, at the same time, there were several nuncios and also the Dalai Lama. The moral of all this is that you should avoid inviting ambassadors and nuncios, if you want to avoid creating a diplomatic incident or starting a religious war.

A further piece of advice we would like to give is to avoid asking a mathematician for age or address at home, if you don't want to be overwhelmed by diophantine equations and more. Around the year 1850 it seems that A. De Morgan[8] was asked what his age was, he replied that:

I will be x after x^2 years.

---

[8]Augustus De Morgan (1806-1871), mathematician and banker founder of Morgan Bank.

The problem is apparently straightforward and indeed it reduces to the second degree equation

$$y + x = x^2 \tag{6.2.1}$$

where y yields the year of birth. The solution of the above equation reads

$$x = \frac{1 + \sqrt{1 + 4y}}{2}. \tag{6.2.2}$$

We do not know y, so we are left with an infinite number of solutions. Age is expressed by integers, we must therefore assume that the delta of the previous solution be a perfect square, namely

$$y = \frac{m^2 - 1}{4}. \tag{6.2.3}$$

Furthermore the delta must also be odd, a condition which yields

$$4y + 1 = (2n + 1)^2 \tag{6.2.4}$$

which in turn yields

$$y = n(n + 1). \tag{6.2.5}$$

By comparing the two results we obtain

$$n = \frac{m - 1}{2} \tag{6.2.6}$$

furthermore, keeping into account that the year when the age has been requested is 1850, we must also have

$$x = 1850 - n(n + 1). \tag{6.2.7}$$

As a consequence n is a number which can be reasonably chosen between $41, 42, 43$. But 41 can be excluded because it yields the year of birth 1722 and therefore in 1850 an age of 128 years, and 43 is excluded because x turns out to be negative. We are therefore left with $n = 42$. We find therefore

$$x = 1850 - 42 \cdot 43 = 1850 - 1806 = 44. \tag{6.2.8}$$

At the time of the question De Morgan was 44. Straightforward!

Ramanujan was asked about his home address and the answer was this:

"I live in a certain street where there are more than 50 and less than 500 houses, the numbering of the house numbers is consecutive on each side (1,2,3 ...). The house number of my home corresponds to a number that is equal to the sum of the house numbers on its right side and those on its left side".

We denote by m the Ramanujan home number and that the total number of houses is n. The conditions of the problem are such that the solution of our problem can be written as

$$1 + 2 + 3 + \cdots + (m - 1) = (m + 1) + (m + 2) + \cdots + n. \quad (6.2.9)$$

If we utilize the notation

$$T_r = \sum_{k=1}^{r} k, \qquad\qquad T_n^{(r+1)} = \sum_{k=r+1}^{n} k, \qquad (6.2.10)$$

we can arrange the first equation in the form

$$\sum_{k=1}^{m-1} k = \sum_{s=m+1}^{n} s \qquad (6.2.11)$$

and eventually write

$$T_{m-1} + T_n^{(m+1)} = T_n. \qquad (6.2.12)$$

A brief intermezzo is necessary, because if we wish to get a solution we must be able to specify what is the final expression for

$$T_r = \sum_{k=1}^{r} k = 1 + 2 + 3 + \cdots + r \qquad (6.2.13)$$

whose derivation can be done by following attribute to a little boy named K. F. Gauss, according to the legend when he was an elementary school pupil[9].

We consider the sum of the first seven natural numbers which are arranged according to the table reported below

$$
\begin{array}{ccccccccc}
1 & 2 & 3 & 4 & 5 & 6 & 7 & + \\
7 & 6 & 5 & 4 & 3 & 2 & 1 & = \\
\hline
8 & 8 & 8 & 8 & 8 & 8 & 8 & = & \frac{7 \cdot 8}{2} & = & 28
\end{array}
\qquad (6.2.14)
$$

We sum the corresponding terms (the last term with the first, the penultimate with the second ...) and we note that the result is always 8. The result we are looking for is therefore half of the product of 7 and 8. Repeating the reasoning for larger series, we arrive at the following formula by induction[10]

$$
T_n = \frac{n(n+1)}{2} \qquad (6.2.15)
$$

Numbers of this type are called *Triangular numbers*. They can indeed be arranged as shown in the Fig. 6.1 where they are displayed as composed by the sum of the red dots, disposed in an equilateral triangle.

Square numbers (namely those arranged in a square) can always be realized as the sum of two complementary triangular numbers

$$
T_m = \frac{m(m+1)}{2}, \qquad T_{m-1} = \frac{(m-1)m}{2} \qquad (6.2.16)
$$

with

$$
T_m + T_{m-1} = m^2. \qquad (6.2.17)
$$

Since the solution of getting the Ramanujan's home number is reduced to the equation

$$
T_{m-1} + T_n^{(m+1)} = \frac{n(n+1)}{2}, \qquad (6.2.18)
$$

---

[9]The agiographers of the great German mathematician K. F. Gauss narrate that he discovered the formula when he was only a junior school pupil. According to the legend the school master assigned the task of summing the first 50 integers and Gauss immediately produced the result in the form $25 \cdot 51$.

[10]This formula justifies also eq. (2.9.2) as the reader can easily prove, if some difficulties arise do not despair and check below.

11

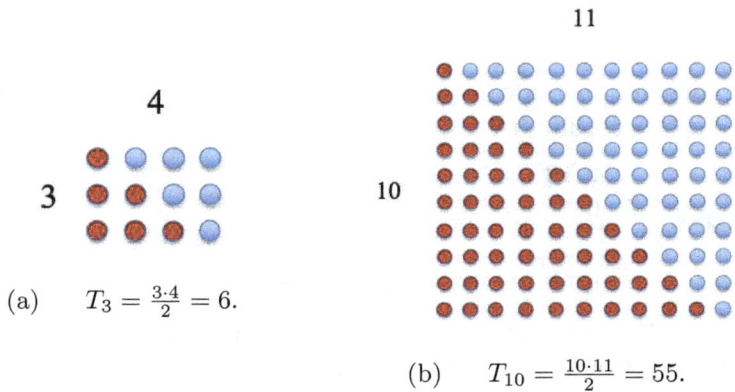

4

3

10

(a) $T_3 = \frac{3 \cdot 4}{2} = 6.$

(b) $T_{10} = \frac{10 \cdot 11}{2} = 55.$

Figure 6.1: Triangular numbers and their geometrical forms.

we are left with

$$m^2 = \frac{n(n+1)}{2}. \tag{6.2.19}$$

The solution of the above problem is to find those numbers (denoted as $_QT_n$) which are either square and triangular. To this aim we rearrange our main equation in the form

$$(2n+1)^2 - 2(2m)^2 = 1. \tag{6.2.20}$$

Setting

$$x = 2n+1, \qquad\qquad y = 2m, \tag{6.2.21}$$

we can reduce our problem to the equation

$$x^2 - 2y^2 = 1 \tag{6.2.22}$$

which is called **equation**[11]. It is evident that the integers numbers generated by the previous identity start from

$$x_0 = 3, \qquad\qquad y_0 = 2. \tag{6.2.23}$$

---

[11] An ancient querelle, raised on the paternity of the Pell equation, attributed it to Euler we must however underline that its origin can be traced back to the 7th century when its study was started by the Indian mathematicians Brahmagupta-Bhaskara.

We make the assumption that all the solutions can be generated as

$$\begin{pmatrix} x_{s+1} \\ y_{s+1} \end{pmatrix} = \begin{pmatrix} x_s & ay_s \\ y_s & x_s \end{pmatrix} \begin{pmatrix} 3 \\ 2 \end{pmatrix} \tag{6.2.24}$$

where a can be fixed by imposing that the s-th solution satisfy the Pell equation. Accordingly we find $a = 2$ thus also stating that the determinant of the matrix is 1. The solution of the above equation is obtained using an extension of the Binet method (according to the recipe discussed later in the chapter), which yields

$$x_s = \frac{\left(3 + 2\sqrt{2}\right)^s + \left(3 - 2\sqrt{2}\right)^s}{2}, \qquad y_s = \frac{\left(3 + 2\sqrt{2}\right)^s - \left(3 - 2\sqrt{2}\right)^s}{2\sqrt{2}} \tag{6.2.25}$$

The solution of our original problem is therefore provided by

$$n_s = \frac{x_s - 1}{2}, \qquad m_s = \frac{y_s}{2}. \tag{6.2.26}$$

The possible results are reported in the Table 6.1.

Table 6.1: **Pell numbers sequence in terms of** s

s	n_s	m_s
1	1	1
2	8	6
3	49	35
4	288	204
5	1681	1189

According to the original constraints ($50 < n < 500$) we find that the Ramanujan's home number is $m = 204$ and the corresponding number of houses is $n = 288$.

    The problem has been solved but wasted effort if we intend to visit Ramanujan, because we do not know the name of the street. Aside from the easy amenities mentioned above, examining a (seemingly) harmless little problem has revealed a world, indeed, absolutely new worlds, which we will try to glimpse further in the next paragraphs.

6.3   Numbers, Polygons and Symbolic Calculus

The discussion and the solution of the problems posed in the previous section, stimulate some thoughts about the meaning of mathematics itself and the infinite ways of "thinking" Mathematics. When asked why one might be interested in a given problem, the answer may simply be "because it was posed". Even if it may sound unprofessional, a mathematicians, with the suitable skills, can have an excellent mathematical production, without any need for it to be included in a strategy that dictates them motivations or specify what should be done and what are the limits of a given discipline.

We will now try to clarify what we want to say. A useful example is the history of Fermat's Last Theorem that could be viewed as a statement devoid of any practical meaning. The notion of practical does not have a precise meaning and is largely misleading if not correctly framed; assuming as practical the application to a physics problem we can say that, to the authors knowledge, this theorem played no role, nor did it seem to have any particular impact on the field of Mathematics itself. For example, Gauss never dealt with it and considered it absolutely irrelevant. Yet it has tormented generations of mathematicians. The search for his demonstration led to construction of an immense building, of which it was only a marginal aspect, alongside an extremely complex theory. Therefore Fermat's Theorem would have had no meaning if it had been formulated as a consequence of the theory of modular forms[12].

In the study of a problem without any academic pretension, i.e. trying to understand where Ramanujan lived, we learned the existence of triangular numbers and following this common thread we could go further. We have introduced a family of numbers as generated by

---

[12]S. Singh, "Fermat's Enigma: The Epic Quest to Solve the World's Greatest Mathematical Problem" Anchor Books (1988).

sums of the type

$$T_n = \sum_{r=1}^{n} r = \frac{n(n+1)}{2}.$$ (6.3.1)

Moreover, as we are curious, we can also ask ourselves what the sum of the consecutive odd numbers is and we would find out that you always get a perfect square, that is

$$\sum_{r=0}^{n-1}(2r+1) = n^2.$$ (6.3.2)

The proof is trivial and, according to what we have learned, is just given in one line, namely

$$\sum_{r=0}^{n-1}(2r+1) = 2T_{n-1} + \sum_{s=0}^{n-1} 1 = 2\frac{(n-1)n}{2} + n = n^2$$ (6.3.3)

and, as we have already underscored, along with triangular, we have discovered square numbers.

We can now proceed further by summing consecutive numbers of the type $3n+1$ and find

$$P_n = \sum_{r=0}^{n-1}(3r+1) = 3T_{n-1} + n = n\frac{3n-1}{2}$$ (6.3.4)

thus getting another family of numbers called *pentagonal numbers*, with the graphical representation reported in Fig. 6.2.

If we proceed further and consider the sum

$$E_n = \sum_{r=0}^{n-1}(4r+1) = 4T_{n-1} + n = n(2n-1)$$ (6.3.5)

we have discovered the *hexagonal numbers*.

We can furthermore establish formal identities of the type

$$P_n = \frac{1}{3}T_{3n-1}, \qquad\qquad E_n = T_{2n-1}$$ (6.3.6)

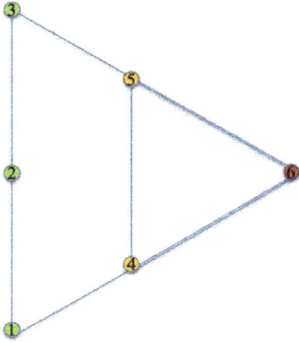

(a) 3° triangular number = 6.

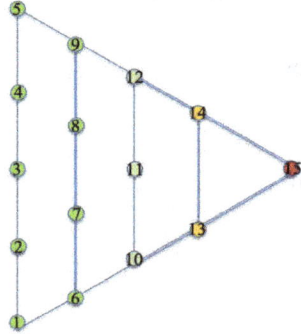

(b) 5° triangular number = 15.

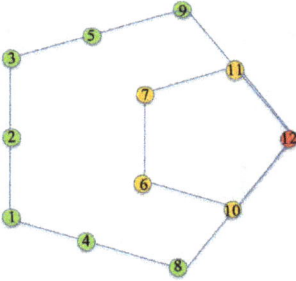

(c) 3° pentagonal number = 12.

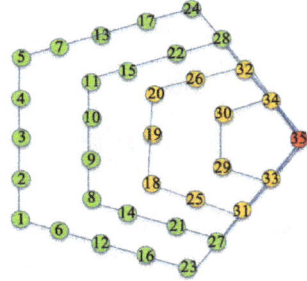

(d) 5° pentagonal number = 35.

Figure 6.2: Polygonal numbers and their geometrical forms.

which allows to consider hexagonal and pentagonal in terms of the triangular family.

The procedure at this point becomes quite sterile (but if one wants to continue having fun in this way why stop him) if we just classify number families. We note however that the search for a number that was triangular and square dragged us towards the Pell equation and the solution of equations to the difference equations of the type

$$x_{s+1} = 3x_s + 4y_s, \qquad\qquad y_{s+1} = 3y_s + 2x_s. \qquad (6.3.7)$$

As for the solution of this system, we hurriedly closed the problem by referring to the generalization of the Binet formula. This is a very vague indication, the method will be illustrated on the basis of an operational formalism. The technique allows to solve, in an elementary way, problems that would be much more involved following a conventional procedure. To this end, we rewrite our system by introducing a mathematical entity, which we will call *displacement operator* defined as follows

$$\hat{E} f_k = f_{k+1}. \qquad (6.3.8)$$

The operator $\hat{E}$ acts on the index k by shifting it by one unit. On the other side if we admit the legitimacy of defining powers of the operator $\hat{E}$, we can define their action on the index k according to the "rule"

$$\hat{E}^m f_k = f_{k+m} \qquad (6.3.9)$$

namely the m-th power of the shift operators produces a jump of m units of the k-index.

We can now use this formal tool to our difference equations as

$$\left(\hat{E} - 3\right) x_s = 4y_s, \qquad\qquad \left(\hat{E} - 3\right) y_s = 2x_s, \qquad (6.3.10)$$

multiplying both sides of the first equation by $\left(\hat{E} - 3\right)$ we end up with

$$\left(\hat{E} - 3\right)\left(\hat{E} - 3\right) x_s = 4\left(\hat{E} - 3\right) y_s. \qquad (6.3.11)$$

Taking into account that for the operators $\hat{E}$ the ordinary algebraic rules are valid, we get

$$\left(\hat{E} - 3\right)\left(\hat{E} - 3\right) = \left(\hat{E} - 3\right)^2 = \hat{E}^2 - 6\hat{E} + 9 \qquad (6.3.12)$$

and after evaluating the action of the single operator on the index s we end up with the difference equation

$$x_{s+2} - 6x_{s+1} + 9x_s = 8x_s \qquad (6.3.13)$$

which can be solved by the use of the Binet method.

A "useful" exercise is that of evaluating the general form of the polygon numbers reported below and study the associated Pell equation.

Table 6.2: **Geometrical interpretation of some families of numbers**

Numbers	General form
Heptagonal	$\frac{1}{2}n(5n - 3)$
Octagonal	$n(3n - 2)$
Ennagonal	$\frac{1}{2}n(7n - 5)$
Decagonal	$n(4n - 3)$

Regarding the pentagonal-square numbers we find, the diofantine equation

$$x^2 - 6y^2 = 1. \qquad (6.3.14)$$

We eventually like to attract the attention on the following relations

$$T_{285} = P_{165} = E_{143} \qquad (6.3.15)$$

and invite the reader to clarify their meaning. We report in Tab. 6.2 the geometrical definition of families of Heptagonal, Octagonal ... numbers.

We mentioned Fermat's Theorem, which was one of a series of theorems that challenged for centuries ranks of professional and amateur mathematicians. Pierre De Fermat, magistrate by profession, was an amateur mathematician. For completeness we report below the theorems that represent Fermat's arithmetic that were the pastimes with which the severe magistrate of the district of Toulouse amused himself to relax after the work, in the winter evenings of the seventeenth century when the calculus took its first steps:

1. each prime number of the form $4n+1$ is the sum of two squares;

2. each prime number of the form $6n + 1$ is the sum of a square and three times a square;

3. each prime number of the form $8n + 1$ and $8n + 3$ is the sum of a square and double of a square;

4. each integer is the sum of three triangular, four square, five pentagonal, six hexagonal, etc.

5. the equation $x^2 - Dy^2 = 1$ always has an integer solution, D being an integer, but not a square;

6. the equations $x^2 + 2 = y^3$ and $x^2 + 4 = y^3$ admit the only integer solutions $(\pm 5, 3)$ and $(\pm 2, 2)$, $(\pm 11, 5)$ respectively.

An innocent problem has brought us so far away, but there is a lot more to say.

6.4 Sums of Numbers, Bernoulli Numbers and Umbral Calculus

Since we have satisfied a certain number of curiosities, let's try to make others come. We will attempt to find out if there is a formula that provides the sum of squares, cubes, fourth powers ... of natural numbers consecutive up to N, that is

$$S_m(N) = \sum_{n=0}^{N} n^m.$$

(6.4.1)

Archimedes had found that[13] (for brevity we omit to report the argument N)

$$S_2 = \frac{2N+1}{3} S_1,$$

(6.4.2)

Nicomacus of Gerasa had found that

$$S_3 = S_1^2.$$

(6.4.3)

The Arabic mathematician Ibn-al-Haitan had shown that

$$S_4 = \frac{S_2}{5}(6S_1 - 1).$$

(6.4.4)

A German mathematician, Johan Falhauber, once known as the great arithmetic of Ulm and who is today unknown even to experts, spent almost all of his life to study sums related to increasing powers, but without being able to classify them all. Jacob Bernoulli (a mathematician that everyone now remembers) took half an hour to calculate them all and he succeeded, proposing a new and amazing family of numbers that today bear his name[14].

Now let's try the following calculation

$$s = 1 + t + t^2 + \cdots + t^N = \sum_{n=0}^{N} t^n.$$

(6.4.5)

The answer is known from elementary algebra texts, but we will propose it again because it is very instructive and useful for future

---

[13] Archimedes obtained this result by the use of the exhaustion method, see R. Netz and W. Noel, "The Archimedes Codes", Phoenix Paperback London 2008.

[14] Jacob Bernoulli in his book Ars Coniectandi defined Faulhaber as the last "Reichmaster" (Arithmetician, Abacus Master ...) and introduced the numbers which now bring his name.

discussion. Multiplying both members of the previous relationship by t gives the following relationship

$$ts = s + t^{N+1} - 1 \tag{6.4.6}$$

which allows us to evaluate the sum as

$$s = \frac{t^{N+1} - 1}{t - 1}. \tag{6.4.7}$$

The use of the same procedure allows to derive the sum

$$S(\vartheta) = \sum_{n=0}^{N} e^{n\vartheta} = \frac{e^{(N+1)\vartheta} - 1}{e^{\vartheta} - 1}. \tag{6.4.8}$$

Why is this result noticeable?

We first note that

1.

$$\lim_{\vartheta \to 0} S(\vartheta) = S_0 = N + 1 \tag{6.4.9}$$

2.

$$\lim_{\vartheta \to 0} \frac{d}{d\vartheta} S(\vartheta) = S_1 \tag{6.4.10}$$

The first identity is a simple consequence of the **De L'Hopital rule**[15], the second is also trivial but important to be detailed for our purposes

$$\frac{d}{d\vartheta} S(\vartheta) = \frac{e^{\vartheta}(1 - e^{N\vartheta}) + N e^{\vartheta(N+1)}(1 - e^{\vartheta})}{(e^{\vartheta} - 1)^2}, \tag{6.4.11}$$

by using again De L'Hopital rule we end up with the result quoted in (6.4.10). A more general result is given below

$$\left(\frac{d}{d\vartheta}\right)^r \sum_{n=0}^{N} e^{n\vartheta}\Big|_{\vartheta=0} = \sum_{n=0}^{N} \left(\frac{d}{d\vartheta}\right)^r e^{n\vartheta}\Big|_{\vartheta=0} = \sum_{n=0}^{N} n^r e^{n\vartheta}\Big|_{\vartheta=0} = S_r(N). \tag{6.4.12}$$

---

[15] We remind that De L'Hopital rule states that if functions $f(x)$, $g(x)$ are such that $\lim_{x \to x_0} f(x) = \lim_{x \to x_0} g(x) = 0$ or $\lim_{x \to x_0} f(x) = \lim_{x \to x_0} g(x) = \infty$ then $\lim_{x \to x_0} \frac{f(x)}{g(x)} = \lim_{x \to x_0} \frac{f'(x)}{g'(x)}$.

The method of the generating function has allowed us to conclude that the sum of r-th powers of the first N integers is "generated" by the function $S(\vartheta)$ according to the identity

$$S(\vartheta) = \sum_{n=0}^{\infty} \frac{S_n(N)\vartheta^n}{n!}. \tag{6.4.13}$$

We established an important result because we managed to tie **Falhauber's sums** in a unitary context.

Let us now try to make some further practical progress. We rewrite the generating function in the following form

$$S(\vartheta) = B(\vartheta)K(\vartheta), \qquad b(\vartheta) = \frac{\vartheta}{e^\vartheta - 1}, \qquad k(\vartheta) = \frac{e^{(N+1)\vartheta} - 1}{\vartheta}. \tag{6.4.14}$$

The function $B(\vartheta)$ is the generating function of **Bernoulli numbers** B_n, according to the formula

$$B(\vartheta) = \sum_{m=0}^{\infty} \frac{B_m}{m!}\vartheta^m \tag{6.4.15}$$

while, as to the function $K(T)$ we find

$$K(\vartheta) = \sum_{r=0}^{\infty}(N+1)^r \frac{\vartheta^r}{(r+1)!}. \tag{6.4.16}$$

After confronting the two series we get[16]

$$S(\vartheta) = \sum_{n=0}^{\infty} C_n \frac{\vartheta^n}{n!}, \qquad C_n = \sum_{r=0}^{n} \binom{n}{r} B_{n-r} \frac{(N+1)^r}{r+1} \tag{6.4.17}$$

and we can conclude that

$$S_n(N) = \sum_{r=0}^{n} \binom{n}{r} B_{n-r} \frac{(N+1)^{r+1}}{r+1}. \tag{6.4.18}$$

---

[16] The product of two series $A(x) = \sum_{r=0}^{\infty} a_r x^r$, $B(x) = \sum_{p=0}^{\infty} b_p x^p$ is treated by means of the so-called **Cauchy product** $A(x)B(x) = \sum_{n=0}^{\infty} c_n x^n$, where $c_n = \sum_{r=0}^{\infty} a_{n-r} b_r$. We remind that $\binom{n}{r} = \frac{n!}{(n-r)!r!}$ denotes the so-called *binomial coefficient*.

How to explicitly calculate B_n numbers? The calculation procedure is simple but tedious. The most direct way is through the series expansion development of the generating function. We can follow an approximate procedure. At the second order in the expansion in ϑ we find

$$S(\vartheta) \simeq \frac{1}{1 + \frac{\vartheta}{2}\left(1 + \frac{\vartheta}{3}\right)} \simeq 1 - \frac{1}{2}\vartheta + \frac{1}{12}\vartheta^2 \tag{6.4.19}$$

which yields

$$B_0 = 1, \qquad\qquad B_1 = -\frac{1}{2}, \qquad\qquad B_2 = \frac{1}{6}. \tag{6.4.20}$$

The iteration of the procedure yields

$$B_m = -\frac{1}{m+1}\sum_{k=0}^{m-1}\binom{m+1}{k}B_k \tag{6.4.21}$$

and it should be noted that all odd index Bernoulli (except $n = 1$) are vanishing. More in general the Bernoulli numbers are given by the identity

$$B_k = \frac{(-1)^k k}{2^k - 1}\sum_{r=1}^{k} 2^{-r}\sum_{j=0}^{r-1}(-1)^j\binom{r-1}{j}(j+1)^{k-1}. \tag{6.4.22}$$

That's all? No and there is something better!!!

What is hidden behind Bernoulli's numbers is something more and perhaps goes beyond what Bernoulli himself could imagine. There is a branch of Mathematics called **Umbral Calculus**. The first time we learned about it we started to fantasize considering a sort of myth of Plato's cave applied to Mathematics. We had like the impression that the mathematical formulas were the "shadows", projected onto some dimension, of deeper forms. It wasn't like that, but we hadn't gone far. The idea of Umbral calculus developed in recent times by Roman and Rota and in more recent times by two of the Authors of this book (Dattoli and Licciardi) was rooted in mathematical conceptions of the aforementioned 19th century operationalists[17].

It was born from the simple observation that sometimes it is convenient, from the computational point of view, treating a series of the type

$$S(x) = \sum_{n=0}^{\infty} \frac{a_n}{n!} x^n \tag{6.4.23}$$

as an exponential function, after replacing a_n with $\hat{a}^n$, namely by promoting the index n to the rank of power exponent. Accordingly we write

$$S(x) = \sum_{n=0}^{\infty} \frac{\hat{a}^n}{n!} x^n = e^{\hat{a}x}. \tag{6.4.24}$$

The procedure can be applied to the Bernoulli numbers thus getting

$$B(\vartheta) = e^{\hat{B}\vartheta} \tag{6.4.25}$$

and furthermore

$$S_n(N) = \sum_{r=0}^{n} \binom{n}{r} \hat{B}^{n-r} \frac{(N+1)^{r+1}}{r+1}. \tag{6.4.26}$$

By treating N as an ordinary variable, we can evaluate the derivative of the function S with respect to N and obtain

$$\frac{d}{dN} S_n(N) = \sum_{r=0}^{n} \binom{n}{r} \hat{B}^{n-r} (N+1)^r = (\hat{B} + N)^n \tag{6.4.27}$$

namely as a Newton binomial. We can even stretch the formalism and note that the integral

$$S_n(N) = \int_0^N (\hat{B} + \tilde{N}) d\tilde{N} \tag{6.4.28}$$

---

[17]The first researches in this context traces back to J. John Blissard, Theory of generic equations, The Quarterly Journal of Pure and Applied Mathematics n. 4, 1861, pp. 279–305. For a modern version see S. Roman, The Umbral Calculus, Pure and Applied Mathematics n. 111, 1984, Academic Press Inc. (Harcourt Brace Jovanovich Publishers), London. For a rigorous treatment of the technique see S. Licciardi, Umbral Calculus: A Different Mathematical Language, Ph.D. thesis, arXiv:1803.03108 [math.CA].

yields the Falhauber sum in the synthetic form

$$S_n(N) = \frac{(\hat{B} + N)^{n+1} - \hat{B}^{n+1}}{n+1}. \tag{6.4.29}$$

We do not know whether it has any meaning for you, but for us it sounded

TERRIFIC!!!

To better appreciate the strength of the formalism and ask the problem of the evaluation of the sum

$$T(N, M) = \sum_{n=0}^{M} S_n(N). \tag{6.4.30}$$

If we had posed this question to Bernoulli he might have provided a quick answer[18]. In his absence we ask the reader.

Before closing this section we make reference to the **Bernoulli polynomials** defined as

$$B_n(x) = (\hat{B} + x)^n = \sum_{s=0}^{n} \binom{n}{s} \hat{B}^{n-s} x^s = \sum_{s=0}^{n} \binom{n}{s} B_{n-s} x^s. \tag{6.4.31}$$

The use of the umbral formalism is extremely advantageous, to state identities like

$$B_n(x + y) = (\hat{B} + x + y)^n = ((\hat{B} + x) + y)^n = \sum_{s=0}^{n} \binom{n}{s} B_{n-s}(x) y^s. \tag{6.4.32}$$

It is now necessary to underscore an important point. The product $\hat{B}^n \hat{B}^m$ does not represent the product of the Bernoulli numbers but should be understood as it follows

$$\hat{B}^n \hat{B}^m = \hat{B}^{n+m} = B_{n+m} \neq B_n B_m. \tag{6.4.33}$$

---

[18] May be not after 1 minute, but certainly he would have come out with a completely new theory in 5 minutes. For a modern view see also G. Dattoli, S. Lorenzutta and C. Cesarano, Finite sums and generalized forms of Bernoulli polynomials, Rendiconti di Matematica, Serie VII Volume 19, Roma (1999), pp. 385–391.

We can e.g. establish the following identity

$$B_n = \frac{1}{2^n}(\hat{B} + \hat{B})^n = \frac{1}{2^n}\sum_{s=0}^{n}\binom{n}{s}B_{n-s}B_s \tag{6.4.34}$$

or the following "duplication Theorem"

$$B_{2n}(x) = \left(\hat{B} + x\right)^n \left(\hat{B} + x\right)^n = \sum_{r=0}^{n}\sum_{s=0}^{n}\binom{n}{r}\binom{n}{s}B_{r+s}x^{2n-r-s}. \tag{6.4.35}$$

Finally it goes by itself that the Bernoulli polynomials are generated by

$$\sum_{n=0}^{\infty}\frac{t^n}{n!}B_n(x) = e^{t(\hat{B}+x)} = \frac{te^{tx}}{e^t - 1}. \tag{6.4.36}$$

We can summarize the message behind the previous formalism by noting that it states the existence of a kind of level of higher abstraction, in which complex forms can be reduced to simple algebraic entities, with the formal rules of monomials (namely $x_n \to x^n$).

In Mathematics there are plethora of families of numbers similar to those of Bernoulli and with analogous properties.

1) The series expansion of the function

$$f(t) = \text{sech}\left(\frac{t}{2}\right) = \frac{2}{e^{\frac{t}{2}} + e^{-\frac{t}{2}}} \tag{6.4.37}$$

can be written as

$$f(t) = \sum_{n=0}^{\infty}\frac{t^n}{n!}E_n, \tag{6.4.38}$$

where E_n are the **Euler numbers**.

2) The series

$$g(t) = \frac{2t}{e^t + 1} = \sum_{n=0}^{\infty}\frac{t^n}{n!}G_n = e^{t\hat{G}} \tag{6.4.39}$$

is used to introduce the **Genocchi numbers**.

The function

$$s(t) = \frac{2t^2}{e^{2t} - 1} \tag{6.4.40}$$

has a corresponding "umbral image" expressed in terms of exponential functions

$$s(t) = e^{t(\hat{B}+\hat{G})}. \tag{6.4.41}$$

These images can be usefully exploited to simplify calculations, The "primitive" of the function $g(t)$ can be formally written as

$$\int_0^t g(\tau)d\tau = \frac{1}{\hat{G}}\left(e^{t\hat{G}} - 1\right). \tag{6.4.42}$$

Playing furthermore with symbols we get e.g.

$$f(t) = \frac{2e^{\frac{t}{2}}}{e^t + 1} = \sum_{n=0}^{\infty} G_n\left(\frac{1}{2}\right)\frac{t^{n-1}}{n!}. \tag{6.4.43}$$

The consequences which can be drawn from the umbral calculus are wide, wild and profound, the examples above are just naïve speculations from a theoretical environment which has opened a sort of revolution, not yet fully accomplished. The implications of the methods go far beyond the aims of this book but it seemed right to provide a mention.

6.5 More Number Families: Ramanujan Quasi-Integers, Euler Lucky, Heegner and Happy Numbers ...

Euler once noted that the trinomial

$$e_l(n) = 2T_{n-1} + 41 = n^2 - n + 41 \tag{6.5.1}$$

reproduces 40 distinct prime numbers for $n \leq 40$. We check indeed that

$$e_l(3) = 47, \qquad\qquad e_l(4) = 53, \qquad\qquad \cdots$$
$$\cdots \qquad\qquad e_l(7) = 83, \qquad\qquad e_l(8) = 97 \tag{6.5.2}$$

which suggests that, although the numbers are primes, they are not all the consecutive primes. It is indeed worth noting that $e_l(7) < 89 < e_l(8)$.

This class of numbers belongs to a more general family called *lucky numbers of Euler*.

The occurrence of such a coincidence is not particularly interesting, there are many of such prime generating polynomials (see below), but we can look at the problem by following a different point of view. We consider the factorization

$$e_l(n) = (n - A_+)(n - A_-), \qquad A_+ = \frac{1 + \sqrt{-163}}{2}$$

(6.5.3)

in which 163 is a prime number, however the previous identity does not hide anything special.

According to what we have learnt on the quadratic irrationals and on the associated Q-Trigonometric functions, we can introduce the sine and cos-like functions

$$_AC(\delta) = \frac{A_+ e^{A-\delta} - A_- e^{A+\delta}}{2\sqrt{-163}}, \qquad _AS(\delta) = \frac{e^{A-\delta} - e^{A+\delta}}{2\sqrt{-163}}$$

(6.5.4)

along with the "complex" exponential

$$_AE(\delta) = e^{A+\delta} = e^{\frac{1}{2}(1+\sqrt{-163})\delta}.$$

(6.5.5)

What does it happen if in place of δ we put π? The result of such a substitution is surprising and we find

$$_AE(-2i\pi) = e^{-i\pi} e^{\pi\sqrt{163}} = -e^{\pi\sqrt{163}}.$$

(6.5.6)

Albeit $e^{\pi\sqrt{163}}$ is probably the most meaningless quantity in the world of numbers, we find that its numerical value is a "quasi-integer"

$$e^{\pi\sqrt{163}} = 262537412640768743.99999999999925\ldots$$

(6.5.7)

which nowadays is called the *Ramanujan constant*.

What is interesting in all this discussion? Probably nothing but, as we have learned, such a question is not properly posed.

Math should also be appreciated for its "experimental" content and facts may be facts, without any special meaning however we like to speculate and therefore we consider the trinomial

$$e_l(n; p) = 2T_{n+1} + p = n^2 + n + p \qquad (6.5.8)$$

(defining the class of lucky numbers of Euler) which is prime generating for $n < 1 < p - 2$ and if $1 - 4p = -h$ where h is a **Heegner numbers** $h = 7, 11, 19, 43, 67, 163$. The $e_l(n; p)$ in terms of **Heegner numbers** read

$$e_l(n; p) = 2T_{n+1} + p = n^2 + n + \frac{h+1}{4} \qquad (6.5.9)$$

which gives rise to the Euler lucky with $p = 2, 3, 5, 11, 17, 41$.

We should add a caveat, the denomination of "***lucky***" has not a meaning in the current sense of the word. The term was coined by *F.L. Lionnais* and the reasons are not clear. Therefore do not waste money (as one of the Authors did) by using them for the power-ball. The reduction of the trinomial $e_l(n; p)$ in quadratic irrationals

$$H_+ = \frac{1 + \sqrt{-h}}{2} \qquad (6.5.10)$$

gives rise to the quasi Ramanujan integers $e^{\pi\sqrt{h}}$.

So what?

To be honest, we do not know. Notwithstanding we can do anything to increase the bewilderment! We propose the following identity for the reader

$$\sqrt[3]{e^{\pi\sqrt{h}} - 744} \simeq N, \qquad\qquad h \geq 19 \qquad (6.5.11)$$

where N is an integer and ask why it is such. If anyone will be able to provide an answer, the e-mails of the Authors are given at the beginning of the book and we would appreciate the comments.

We can do something more, to throw further discredit on the world of mathematicians by introducing **happy and unhappy numbers**.

We consider the number 19 and note that

$$19 \to 1^2 + 9^2 = 82 \to 8^2 + 2^2 = 68 \to 6^2 + 8^2 = 100 \to 1^2 + 0^2 + 0^2 = 1 \tag{6.5.12}$$

and conclude that 19 is *happy*. Not clear? If so we take 23 and find

$$23 \to 2^2 + 3^2 = 13 \to 1^2 + 3^2 = 10 \to 1^2 + 0^2 = 1 \tag{6.5.13}$$

so 23 is happy! The same conclusion holds for 28

$$28 \to 2^2 + 8^2 = 68 \to 6^2 + 8^2 = 100 \to 1^2 + 0^2 + 0^2 = 1. \tag{6.5.14}$$

Obviously a number is happy if the procedure of adding the squares of the digits that compose it and of those composing the numbers obtained through this procedure ends with 1. *Unhappy* are those numbers that do not satisfy this condition and dramatically unhappy are those who after a few iterations of the above type reproduce themselves (for more specific examples see *OEIS* A031177).

The algorithm for the definition of happy (or unhappy) can be extended to the sum of the cube of the digits. Within this context it can be checked how much is unhappy 133!

$$133 \to 1^3 + 3^3 + 3^3 = 55 \to 5^3 + 5^3 = 250 \to 2^3 + 5^3 + 0^3 = 133 \ldots . \tag{6.5.15}$$

Just to end we quote the *perfect numbers* like 6, 28 which is equal to the sum of its divisors

$$6 = 1 + 2 + 3, \qquad\qquad 28 = 1 + 2 + 4 + 7 + 14. \tag{6.5.16}$$

What about **amicable numbers**?

They are couple of numbers in which one is equal to the sum of the divisors of the other and vice-versa. The reader may check that

220 and 284 are such.

About *Fiancé numbers*?

Couple of numbers similar to amicable in which the sum of the divisors excluding 1 is equal to the other and vice-versa (check 48 and 75).

And about *Abundant numbers*?

Number which are less than the sum of their divisors. Number 12 is abundant, we have indeed

$$1 + 2 + 3 + 4 + 6 = 16 \tag{6.5.17}$$

with an abundance of 4.

That's enough! Let us move towards less exoteric fields.

6.6   Diophantine Equations and Continuous Fractions

Diophantus of Alexandria lived between the third and fourth centuries after Christ and the duration of his life can be deduced from the following problem reported as an epitaph on his grave:

His youth lasted 1/6 of his life;
then his beard started to grow after 1/12;
he married after 1/7 and was born a son after 5 years.
The son lived the father's mid years
and the father died 4 years after the son.
How long did Diophantus live?

This is certainly an original and concise way to certify the important steps of an existence and to certify his own departure. The problem he posed to certify his "CV" was tremendous for the mathematical literacy of the time. Diophantus was the first to deal with equations in several variables that admit solutions only in terms of integers.

Problems of this type have been touched on several times during this and previous chapters. The equation of Pell is an example of diophantine equation, as well as the problem of asking whether the equations

$$2x+3y = 11, \qquad 7x^2-5y^2+2x+4y-1 = 0, \qquad y^3+x^3 = z^3$$
$$(6.6.1)$$

admit solutions in the domain of integers. Apart from the linear case, the problem it is far from being trivial.

Hilbert's tenth problem, one of the 23 proposed by him in 1900 and which would have led the development of the mathematics of the twentieth century, concerned precisely the request for a "universal" method for solving diophantine equations. The answer came 70 years later and in fact Mateyasivich showed that such a possibility does not hold. Said in less approximate terms the Theorem sounds as follows:

"Given a generic diophantine equation, there is no algorithm that can predict whether such an equation has a solution or not."

Note that the Theorem refers to "algorithm" a concept that must be integrated with the notion of "Computability", introduced by Gödel, Kleene and Turing in the 1930s. We think it appropriate to spend two words on this topic, perfectly on line with the previous discussion.

The solvability of a problem is linked to its computability which is in turn linked to a "machine", namely to a formalism, capable of calculating functions associated with a given problem. Turing has defined one appropriate machine (the *"Turing machine"*) which computes the function associated with a certain problem in a finite number of steps. If this is not possible the function is not computable and therefore the problem is not solvable or as they say in jargon is not decidable.

After this dutiful reference we move on to a specific problem, which involves the solution of a diophantine linear equation:

A society has decided to invest 810.000 dollars to build a certain number of houses by spending 80.000 dollars for each and some sheds at 50.000 dollars a piece how many apartments and sheds will be built?

If we use x to denote the number of houses and y that of sheds, we find

$$80000x + 50000y = 810000 \qquad (6.6.2)$$

or

$$8x + 5y = 81 \implies y = -\frac{8}{5}x + \frac{81}{5}. \qquad (6.6.3)$$

If we had not any limitations, all the pairs of numbers (x, y) satisfying the condition $y = -\frac{8}{5}x + \frac{81}{5}$ would be acceptable. Obviously the only possibilities are pairs of integers and non-negative numbers, in our case $(2, 13)$ and $(7, 5)$ (see Fig. 6.3).

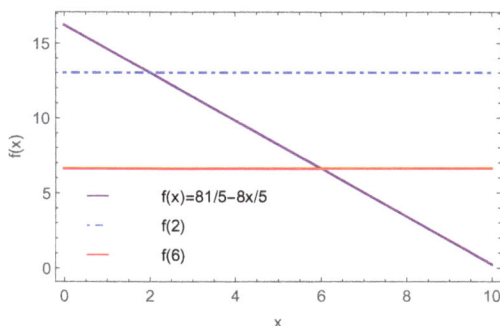

Figure 6.3: Solution of equation $8 \cdot 10^4 x + 5 \cdot 10^4 y = 81 \cdot 10^4$.

We have found the solution to the previous problem by the use of a naïve method employing a geometrical support, a more general method could certainly be more effective for solving a whole class of problems.

Let's see now how we can write a solution in less limited terms. We consider therefore the equation

$$mx + ny = p, \qquad \forall m, n, p \in \mathbb{N} \qquad (6.6.4)$$

where x and y are the unknown of our problem with the restriction that they are integers. We indicate with the symbol (a, b) the greatest common divisor (mcd) between the numbers a and b. We will say that our equation admits infinite solutions if the inhomogeneous term p of the equation is divisible by the greatest common divisor of the linear coefficients. The first step towards the solution is the search for a particular solution which we will write as follows

$$x_0 = \frac{rc}{(m, n)}, \qquad\qquad y_0 = \frac{sc}{(m, n)} \qquad (6.6.5)$$

with r, s any two positive integers. All the other solutions are linked to these by

$$x = x_0 + n\frac{k}{(m, n)}, \qquad y = y_0 + m\frac{k}{(m, n)}, \qquad k = 0, \pm 1, \pm 2, \pm 3, \ldots$$
$$(6.6.6)$$

It is not difficult to establish the correctness of the solutions with respect to the previous problem, which, while admitting infinite whole solutions, it has limitations related to positivity and absolute values.

The problem could be solved in even more general terms therefore consider the equation

$$ax + by = c \qquad (6.6.7)$$

with a, b integers. We use the following expansion in terms of *continued fractions*[19]

$$\frac{a}{b} = [a_1; a_2, \ldots, a_n] \qquad (6.6.8)$$

---

[19]C. D. Olds, Continued fractions, Mathematical Association of America, January 2012, print publication year: 1963, https://doi.org/10.5948/UPO9780883859261.

along with the reduced forms p_n, q_n thus getting the particular solution

$$x_0 = p_{n-1}, \qquad\qquad y_0 = q_{n-1}. \qquad (6.6.9)$$

All the others are expressible as

$$x = c\, q_{n-1} - kb, \qquad\qquad y = -c\, p_{n-1} + ak \qquad (6.6.10)$$

and the relevant proof is straightforward. For completeness sake, we like to add that the previous identity is an effective solution of our problem. The use of the relationship

$$a\, q_{n-1} - b\, p_{n-1} = 1 \qquad (6.6.11)$$

allows the straightforward conclusion

$$a(c\, q_{n-1} - kb) + b(-c\, p_{n-1} + ak) = 1 \;\Rightarrow\; c(a\, q_{n-1} - b\, p_{n-1}) = 1. \qquad (6.6.12)$$

The following problem has historical roots[20] and we have formulated it in more modern terms.

A gang with five components has earned through their illegal activities a certain amount of millions of dollars and have the agreement that it will be divided equally among them. They do not know the total amount. This is not a gentlemen's agreement and at night one of the protagonists wakes up and divides the swag in five equal piles of millions of dollars and discovers that one million has advanced. Since he is wise and provident, takes the part due and also the million Euro more for creating funds for pension. A second component has the same idea, wakes up: divides in five equal piles the remaining parts of millions, realizes that there is a surplus of one million and proceeds like his colleague. So do the remaining gang members. The next morning they divide what is left, which a number of millions divisible by 5. What is the total budget originally "saved" by the five?

The solution is amusing and not particularly complicated.

---

[20]See the already quoted book Continued Fractions, the problem originally referred to Sailor Men, Monkeys and Coconuts.

We note that the last leaves a sum which is a multiple of 5, let's say $5y$, it means that the amount of money left by his colleague is $S_1 = 5(5y) + 1$. By iterating the procedure we eventually find the recursion

$$S_{n+1} = 5S_n + 1, \qquad\qquad S_0 = 5y. \qquad\qquad (6.6.13)$$

Now assuming $y = 1$ we get that

$$S_1 = 26, \quad S_2 = 131, \quad S_3 = 656, \quad S_4 = 3281, \quad S_5 = 16406. \quad (6.6.14)$$

The initial amount of millions is therefore 16406. The answer is not given in absolute, it depends on the value of y which has been arbitrarily assumed to be 1. A different solution including the dependence on y is

$$S_5 = 5^5 y + \frac{5^5 - 1}{4} \qquad\qquad (6.6.15)$$

and the reader is invited to prove it.

6.7 Repeated Radicals

We walked away quite simply following the thread of our thoughts and we did a little tour starting from a problem that appeared in a newspaper of popular puzzles.

Let's go back to Ramanujan, the supposed subject of this chapter. In the aforementioned article by Cartier, Ramanujan is ascribed to a very particular species of mathematicians that he calls *mathemagicians*, whose physiognomy is difficult to define, but which is not difficult to imagine. The feature the main reason for this genius is to derive a result that is anything but from an absolutely banal relationship evident and ... vice versa. Most of us have no difficulty accepting the fact that

$$x = \sqrt{x \sqrt{x^2}}. \qquad\qquad (6.7.1)$$

Its iteration in terms of repeated radicals (namely by replacing compulsively x^2 with $x\sqrt{x^2}$, we find

$$x = \sqrt{x\sqrt{x\sqrt{x\sqrt{x\sqrt{x}\ldots}}}}$$ (6.7.2)

which yields for example that

$$2 = \sqrt{2\sqrt{2\sqrt{2\sqrt{2\sqrt{2}\ldots}}}}$$ (6.7.3)

is not a completely trivial identity, as that reported in the starting example.

If we push the idea we have tried to convey with the previous examples, we can write the further iterate radical identity

$$x = \sqrt[3]{x^2\sqrt[3]{x^3}} = \sqrt[3]{x^2\sqrt[3]{x^2\sqrt[3]{x^2}\ldots}}$$ (6.7.4)

We can safely conclude

$$2 = \sqrt[3]{4\sqrt[3]{4\sqrt[3]{4\sqrt[3]{4}\ldots}}}$$ (6.7.5)

The following identity is evident

$$x^2 = x + x(x-1).$$ (6.7.6)

Furthermore limiting ourselves to positive numbers, we find

$$x = \sqrt{x + x(x-1)}$$ (6.7.7)

and then

$$x = \sqrt{x + x(x-1)\sqrt{x + x(x-1)\sqrt{x + \ldots}}}$$ (6.7.8)

which provide the relationships

$$2 = \sqrt{2 + \sqrt{2 + \sqrt{2 + \sqrt{2 + \ldots}}}},$$

$$3 = \sqrt{3 + \sqrt{3 + \sqrt{3 + \sqrt{3 + \ldots}}}}, \qquad (6.7.9)$$

$$4 = \sqrt{4 + \sqrt{4 + \sqrt{4 + \sqrt{4 + \ldots}}}}$$

$\ldots$

It is also true that

$$x = \sqrt[3]{x + x(x^2 - 1)} \qquad (6.7.10)$$

which eventually yields

$$2 = \sqrt[3]{2 + 3\sqrt[3]{2 + 3\sqrt[3]{2 + 3\sqrt[3]{2 + \ldots}}}} \qquad (6.7.11)$$

and, as a particular case, the identity

$$2 = \sqrt[n]{2 + (2^{n-1} - 1)\sqrt[n]{2 + (2^{n-1} - 1)\sqrt[n]{2 + (2^{n-1} - 1)\sqrt[n]{2 + \ldots}}}} \qquad (6.7.12)$$

which is an identity slightly more complicated than those stated at the beginning of this section.

Let us now consider the relationship

$$n(n + 2) = n\sqrt{1 + (n + 1)(n + 3)} \qquad (6.7.13)$$

and introduce

$$f_n = n(n + 2) \qquad (6.7.14)$$

which allows the recursion

$$f_n = n\sqrt{1 + f_{n+1}} \qquad (6.7.15)$$

which iterated with repeated radical finally yields

$$f_n = n\sqrt{1+(n+1)\sqrt{1+f_{n+2}}} = n\sqrt{1+(n+1)\sqrt{1+(n+2)\sqrt{1+f_{n+3}}}} = \ldots \tag{6.7.16}$$

and since $f_1 = 3$ we end up with

$$3 = \sqrt{1+2\sqrt{1+3\sqrt{1+4\sqrt{1+\ldots}}}} \tag{6.7.17}$$

which is the question raised at the beginning of this chapter.

A fairly natural result within the present context, but... it is matter of being Ramanujan to catch the answer out of any context.

We have learned a method and we are now able to produce results of the same type at industrial level. If we keep the square of both sides of the previous identity we find

$$9 = 1+2\sqrt{1+3\sqrt{1+4\sqrt{1+\ldots}}} \rightarrow 4 = \sqrt{1+3\sqrt{1+4\sqrt{1+5\sqrt{1+\ldots}}}} \tag{6.7.18}$$

and even

$$5 = \sqrt{1+4\sqrt{1+5\sqrt{1+6\sqrt{1+\ldots}}}} \tag{6.7.19}$$

Let us now consider the identity

$$4 = \sqrt{6+2\sqrt{7+3\sqrt{8+4\sqrt{9+\ldots}}}} \tag{6.7.20}$$

which we assume given for granted. Let us proceed as indicated before and find

$$5 = \sqrt{7+3\sqrt{8+4\sqrt{9+5\sqrt{10+\ldots}}}},$$

$$6 = \sqrt{8+4\sqrt{9+5\sqrt{10+6\sqrt{11+\ldots}}}}, \tag{6.7.21}$$

$\ldots$

which, at the end, suggest the general form from which all the previous identities can be derived

$$g_n = n\sqrt{(n+5)} + g_{n+1}, \qquad\qquad g_n = n(n+3). \qquad (6.7.22)$$

The procedure may lead much further, if we set

$$f(x) = x + n + a \qquad\qquad (6.7.23)$$

the identity[21]

$$f(x) = \left[ax + (n+a)^2 + xf(x+n)\right]^{\frac{1}{2}} \qquad (6.7.24)$$

holds and by iterating the procedure we can obtain

$$x + n + a = \sqrt{ax + (n+a)^2 + x\sqrt{a(x+n) + (n+a)^2 + (x+n)\sqrt{A}}}$$
$$A = a(x+2n) + (n+a)^2 + (x+2n)\sqrt{\ldots}$$
$$(6.7.25)$$

which offers the opportunity of deriving a plethora of "strange" identities. Setting e.g. $x = n = a = 1$ the previous identity yields

$$3 = \sqrt{3 + \sqrt{4 + 2\sqrt{5 + 3\sqrt{\ldots}}}} \qquad (6.7.26)$$

We invite the reader to keep notice of the recurrence

$$f_m(x_1, x_2 + x_3) = \sqrt{x_2 x_1 + (x_2 + x_3)^2 + x_1 f_{m-1}(x_1 + x_3, x_2 + x_3)},$$
$$f_0(a, b) = a + b$$
$$(6.7.27)$$

and draw suitable consequences.

This is perhaps enough but, if you like, you may proceed further. Anyone gets fun in the way he/she prefers, nobody is going to judge you for your arithmetic preferences.

---

[21]Bruce C. Berndt, Ramanujan's Notebook, 1989, Part II, p. 108, Springer Verlag.

6.8 Madness as a Method of Investigation

At the end of this path we can safely say that prejudices do not have in Mathematics any reason to exist!

We have seen how the prejudice that square roots of negative numbers did not exist has been largely overcome by the introduction of a class of numbers called, unhappily or happily, imaginary. We also know that the sum of the internal angles of a triangle is not always a flat angle and that the idea of the parallelism between two straight lines is certainly more complicated than Euclid had initially suggested. Finally, the last century was shaken by Gödel's proof that there are truths that cannot be demonstrate as such and that two conflicting "truths" can also exist.

We will see later that there is no need to have prejudices against divergent series and that we should not believe that $\sum_{n=1}^{\infty} n = $ is an infinite number.

It is well known that to overcome prejudices one must broaden his own point of view, the same thing happens in the context of mathematics. The ability to enlarge your horizon and accept new concepts allows you to jump conceptual and leaps forward that would not have been possible if they had risen up fences, as they did at the time when Pythagoreans banished irrational numbers from the domain of mathematical harmony.

Now let's try to give an example of how changing a point of view a totally new world can emerge.

Example 19. *We know that*

$$\int x^n dx = \frac{1}{n+1} x^{n+1} + c. \tag{6.8.1}$$

If we critically examine it, we are faced with a problem: in the case $n = -1$ *we get an indeterminate form, which does not allow the definition of the function associated with the integral of* x^{-1}. *Since we are faced with an apparent divergence, we try to "subtract" the*

*infinite, which derives from the first term, through an appropriate
choice of the arbitrary constant c. For example, if we write*

$$c = -\frac{1}{n+1} + d \;\Rightarrow\; \int x^{-1}dx = \lim_{n\to-1}\frac{1}{n+1}\left(x^{n+1}-1\right)+d \quad (6.8.2)$$

we get

$$\lim_{n\to-1}\frac{1}{n+1}\left(x^{n+1}-1\right) = \lim_{n\to-1}\frac{1}{n+1}\left(e^{(n+1)\ln(x)}-1\right) = \ln(x)$$
$$(6.8.3)$$

and therefore

$$\int x^{-1}dx = \ln(x) + d \qquad\qquad (6.8.4)$$

as we know from elementary calculus[22].

What have we done? We have brought out the logarithm function
by eliminating an infinite that masked its existence, through a suitable technique of *subtraction (of infinities)*. The procedure for subtracting infinities, called *renormalization*, is more known to physicists
than to mathematicians and constitutes one of the elements of the
most heated discussion between rigorist and their opponents. Even
some physicists have been looking at its legitimacy suspiciously[23].

Pushing further these ideas it is possible to prove that[24]

$$\sum_{n=0}^{\infty} n = 1+2+3+4+\cdots = -\frac{1}{12}, \qquad \sum_{n=0}^{\infty} n^3 = 1+2+3+4+\cdots = \frac{1}{120},$$
$$(6.8.5)$$

---

[22]This argument had been suggested to one of the Author (G. D.) by T. E. O.
Hermsen.

[23]See also G. Dattoli and M. Del Franco, The Euler Legacy to Modern Science,
Lecture Notes of Seminario Interdisciplinare di Matematica, vol. 9, 2010, pp.
1–24.

[24]To be more precise we should write $\sum_{n=0}^{\infty} n = 1 + 2 + 3 + 4 + \cdots = \left(-\frac{1}{12}\right)_R$
where the subscript R indicates that a subtraction procedure has been adopted.
For an introduction to the Theory of Normalization see E. Elizalde and A. Romeo
"Zeta Regularizations with Applications" World Scientific (1994) and B. Delam-
otte "A hint to renormalization" Am. J. Phys. 72 (2004) 170. For an introduction
to Casimir effect see A. Lambrecht, "The Casimir effect: a force from nothing",
Physics World, September 2002.

and even

$$\infty! = \sqrt{2\pi} \; . \tag{6.8.6}$$

Relations of this type despite their "unbelievability" become fundamental for the calculation of measurable quantities in the field of physical problems. Ramanujan (before him Euler and Abel) was one of the first to discover the world of divergent series and to build a theory on it on which we cannot dwell.

But let's see what the possible relapses are, analyzing the so-called **Casimir effect**, shown in the Fig. 6.4 which shows two conductive surfaces, facing each other and placed at a distance alpha attract each other due to quantum fluctuations.

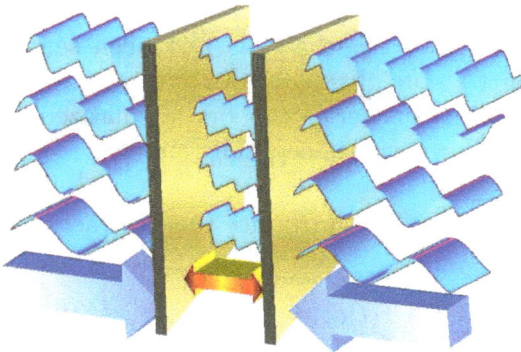

Figure 6.4: Conducting plates and Casimir effect.

From a physical point of view it can be said that the external vacuum radiation is greater than the internal one (that is minor, due to the confinement conditions of the radiation itself) and the plates are pushed towards one to the other by the radiation pressure. In formulas we will have

$$F = -\frac{\partial}{\partial a} \frac{\langle E \rangle}{A} \tag{6.8.7}$$

where $\frac{\langle E \rangle}{A}$ is the confined energy per unit surface. The explicit computation yields (if you ask why look at the literature quoted at the end of the chapter)

$$\left|\frac{\langle E \rangle}{A}\right| = \frac{\hbar c \pi}{6a^3} \sum_{n=0}^{\infty} n^3. \tag{6.8.8}$$

People with "esprit de geometrie"[25] would conclude that something went wrong or that the starting point is false. Others with esprit de finess miht apply the Ramanujan sum and find

$$\left|\frac{\langle E \rangle}{A}\right| = \frac{\hbar c \pi}{720a^3} \tag{6.8.9}$$

which is compatible with the experimental results.

Pure chance, Madness ... or whatever else. May be!

To conclude and to tie what has just been said with the previous discussion we note that the trick is that of subtracting from an infinite and divergent series can be "regularized, domesticated, renormalized... following a conceptually simple procedure.

For any diverging series, we introduce the subtraction criterion

$$\sum_{n=0}^{\infty} f(n) - \int_0^{\infty} f(t)dt = -\sum_{k=1}^{\infty} \frac{B_k}{k!} f^{(k-1)}(0) \tag{6.8.10}$$

where B_n are the Bernoulli numbers and $f^{(m)}(0)$ the m-th derivatives of $f(x)$ calculated at $x = 0$. As a particular case we find

$$\sum_{n=0}^{\infty} n - \int_0^{\infty} t \, dt = -B_1 f(0) - B_2 \frac{f^{(1)}(0)}{2!} = -\frac{1}{12}. \tag{6.8.11}$$

Finite and infinite sums of the type $\sum_{n=0}^{\infty} n^p$ are linked to the Bernoulli numbers and to their properties. Since we have mentioned

---

[25]Blais Pascal introduced the concepts of "esprit de geometrie" and "esprit de finess" people possessing the second characteristic are able of understanding paradoxical novelties.

the masses of quarks, we like to quote an empirical **formula** due to **Barut**[26]

$$M(N) = M_e \left[1 + \frac{2}{3}\alpha^{-1} \sum_{n=0}^{N} n^4 \right] \qquad (6.8.12)$$

where M_e is the electron mass and α the fine structure constant. Using what we have learnt so far, we can conclude that

$$M(N) = M_e \left[1 + \frac{2}{3}\alpha^{-1} \sum_{r=0}^{4} \binom{4}{r} B_{4-r} \frac{(N+1)^{r+1}}{r+1} \right]. \qquad (6.8.13)$$

In the previous relation N represents the order of the lepton family, $N = 0$ corresponding to e and the other to $(\mu, \tau, \dots)$ and if the families where infinite we would find

$$M(\infty) = M_e \left[1 - \frac{2}{3}\alpha^{-1} B_4 \right] \qquad (6.8.14)$$

... as to the last crazy identity formula $\infty! = \sqrt{2\pi}$? ... Just a hint:

$$\infty! = \lim_{n \to \infty} 1 \cdot 2 \cdot 3 \cdot \dots \cdot n = e^{\left(\sum_{n=0}^{\infty} \ln(n)\right)} = \dots \qquad (6.8.15)$$

6.9 Final Comments: Sir Winston Churchill and Mordechai Shapiro

Two "universal" constants $c, \hbar$ (light velocity and reduced Planck constant respectively) have been used in the Casimir energy formula. There is the unconfessed dream, which crosses the world of Physicists (and not only), that it is possible to reduce the universal constants to relations between numbers with particular "properties", be they prime, irrational or, even better if transcendental numbers[27]. It is a periodically recurring alchemical dream[28], neither were the great

---

[26] A. O. Barut, Lepton mass formula, Phys. Rev. Lett. n. 42, 1979, p. 1251.

[27] A. I. Miller, Deciphering the Cosmic Number, Deciphering the Cosmic Number (137): Jung, Pauli, and the Pursuit of Scientific Obsession W. W. Norton & Co. (2009) ISBN 0-393-06532-4.

[28] G. Dattoli, The fine structure constant and the numerical alchemy, arXiv:1009.1711 [physics.gen-ph], http://arxiv.org/ftp/arxiv/papers/1009/1009.1711.pdf.

minds immune.

The value of the fine structure constant, or the quantity that regulates the "intensity" of the electromagnetic interactions, for example, it is reproduced through **Gilson's formula**

$$\alpha = \frac{\pi}{\beta} \cos(\beta) \tanc\left(\frac{\beta}{29}\right), \qquad \tanc(x) = \frac{\tan(x)}{x}, \qquad \beta = \frac{\pi}{137}.$$
(6.9.1)

An alchemical formula, in the sense mentioned above, because it is based on 2 prime numbers $(137, 29)$ and a transcendental number. Although the formula provides a value $(137.035999786699\ldots)$ in agreement up to the eleventh decimal place with the experimental value, it is difficult to ascribe to it any scientific value. Yet the neo-plato-Pythagoreans have no qualms and are able to profit from any combination of numbers.

The Pythagorean triad is used to suggest esoteric connections with the speed of light, in fact

$$\frac{c}{10^8} = 2.99792458 \simeq \sqrt{3\sqrt{4\sqrt{5}}}.$$
(6.9.2)

Furthermore the use of constants like

$$s = 2 + 2\cos\left(\frac{2\pi}{7}\right)$$
(6.9.3)

representing the length of the size of an eptagon inscribed in a unit circle of radius 1, does not play a minor role, in providing

a) The light velocity

$$c \simeq \left[\frac{30(9s+8)}{67s+2}\right]^{12} = 2.99792457.$$
(6.9.4)

b) The gravitational constant

$$G \simeq \left[7 + \frac{s}{70}\right]^{-12} \simeq 6.67426 \cdot 10^{-11}$$
(6.9.5)

with $6.67426 \cdot 10^{-11}$ experimental value.

c) The **Planck constant**

$$2\pi\hbar \simeq \left[\frac{s^3}{17 + 40\pi^2\sqrt{70}}\right]^{-12} \simeq 6.626068966 \cdot 10^{-34} \qquad (6.9.6)$$

with $6.626068966 \cdot 10^{-34}$ experimental value.

In 1654 **Cartesius** formulated the so-called *Kissing Circles Theorem*

$$\left(\chi_A^2 + \chi_B^2 + \chi_C^2 + \chi_D^2\right) = 2\left(\chi_A + \chi_B + \chi_C + \chi_D\right)^2 \qquad (6.9.7)$$

in which χ are the curvatures $\left(\frac{1}{r}\right)$ of the circles arranged in one of the configurations (named after Cartesius) reported in Fig. 6.5.

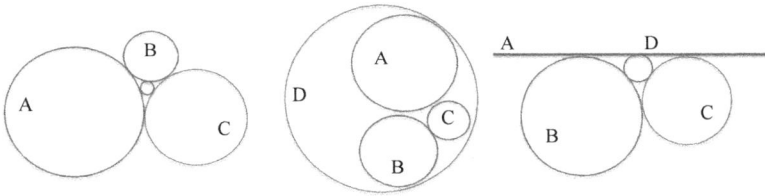

Figure 6.5: Cartesius Theorem.

The theorem fully respectable, within a mathematical context, has been adapted to high energy physics to suggest the following empirical form[29]

$$\left(m_e + m_\mu + m_\tau\right) = \frac{2}{3}\left(\sqrt{m_e} + \sqrt{m_\mu} + \sqrt{m_\tau}\right)^2 \qquad (6.9.8)$$

where m_e, m_μ, m_τ are the masses of leptons. Although general relativity might legitimate the idea of associating a curvature with the

---

[29]The discrepancy with the proportionality factors $\left(2, \frac{2}{3}\right)$ had been discussed by J. Kocik, The Koide mass formula and geometry of circles, arXiv:1201.2067 [physics.gen-ph], http://arxiv.org/pdf/1201.2067.pdf and A. Rivero and A. Gsponer, The strange formula of Dr. Koide, 2008, arXiv:hep-ph/0505220v1 25 May 2005.

masses, the previous formula is scarcely reliable within the trend of high energy physics.

We must furthermore underline that an analogous coincidence holds for quark masses too. We get indeed

$$\frac{m_d + m_u + m_s}{\left(\sqrt{m_d} + \sqrt{m_u} + \sqrt{m_s}\right)^2} = \frac{5}{9}. \tag{6.9.9}$$

The use of the definition of the quark masses in terms of the golden ratio (see eqs. (3.5.3)) agrees very well with such a relationship.

Before closing these comments, we like to make a final remark, by underscoring that another quantity associated of crucial importance in high energy Physics can be written in terms of the ϕ golden ratio[30].

The Cabibbo angle is a physical constant controlling specific decay processes in high energy Physics. It is usually expressed through the identity[31] $\theta_c \simeq \sqrt{\frac{m_d}{m_s}}$, which in terms of ϕ eventually writes

$$\theta_c \simeq \phi\sqrt{\frac{\phi}{5}} \simeq 0.216. \tag{6.9.10}$$

For those who like the geometrical fascination, we have reported in Fig. 6.6 the geometrical construction of the golden spiral, along with the golden rectangle and successive powers of ϕ^{-1}.

The idea of adjusting masses of fundamental entities in such a regular environment is certainly endearing, but very hard to be reconciled with the view tracing the origin of the masses to the Higgs mechanism.

We are at the end of this journey in Mathematics. The authors wanted to transmit among others things the idea that it can be fun

---

[30]Y. Kajiyama, M. Raidal and A. Strumia, Golden ratio prediction for solar neutrino mixing, Phys. Rev. D, 76, 2007, 117301.

[31]C. D. Froggatt and H. B. Nielsen, Hierarchy of quark masses, Cabibbo angles and CP violation, Nuclear Physics, B 147(3), 1979, pp. 277–298.

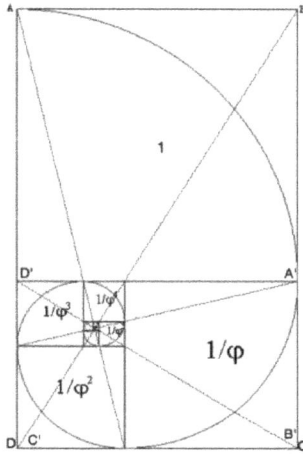

Figure 6.6: Golden spiral and segment associated with the inverse of the golden ratio.

to play with numbers and retrieve old computational methods that shaped both the History of Mathematics and the civilizations themselves.

However, we deem it necessary to insist on alleged esoteric properties of numbers as it can become dangerous and considering numbers as a key to deciphering a pre-ordered code can become an unjustified and dangerous waste of time.

We can, in complete peace of mind, assert that the relationship between the Pythagorean triples and the speed of light has no more physical meaning than the geometric observation reported in the Fig. 6.7, which shows how in a "Egyptian" Pythagorean triangle the diameter of the circles tangent to the inscribed circle and the base are the inverse of the square of the silver ratio and the golden ratio[32].

At the same time we do not deny that the numbers have something *apparently magic*, they are however human "artifacts" albeit the most... elusive.

---

[32] J. Kapusta, The Square, the Circle and the Golden Proportion: A New Class of Geometrical Construction, Forma n. 19, 2004, pp. 293–313.

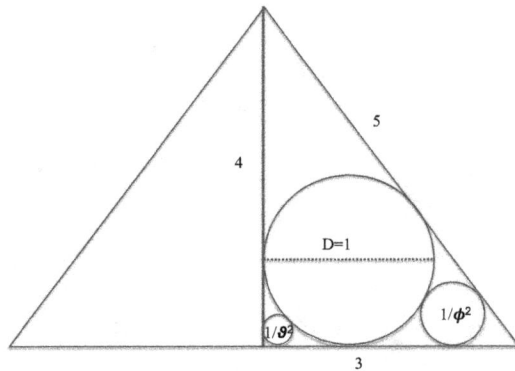

Figure 6.7: Egyptian triangle and inscribed circles. The circle inscribed within half of the triangle has a diameter, $D = \frac{1}{3}$ (rescaled to 1). Two circlestangent to the unit circle and the base are inverse squares of the golden and silver proportions.

The translation of their properties in geometrical terms possesses un-dubitable fascinations, like that associated with the identity

$$\sum_{n=2}^{\infty} \varphi^{-n} = 1 \qquad (6.9.11)$$

whose geometrical interpretation is given in Fig. 6.8.

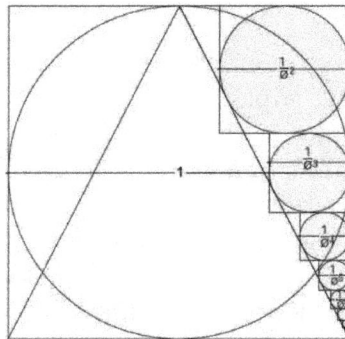

Figure 6.8: Eq. (6.9.11) and its geometrical interpretation.

The following figure shows an artistic interpretation of Fibonacci numbers and Pythagoras' Theorem.

Yes it deals with beautiful and amusing things[33], but ascribing to these any hidden meaning is certainly useless Exercise.

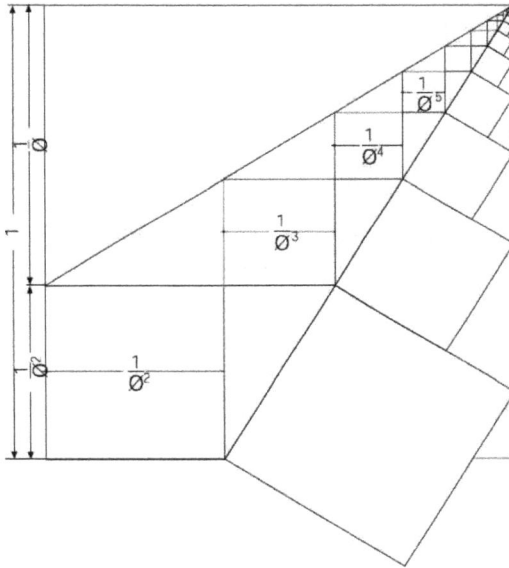

Figure 6.9: Pythagoras' Theorem and Fibonacci Numbers.

---

[33] See the already quoted essay by Janusz Kapusta.

We would like to conclude this ending Chapter by making reference to two different points of view. The first by *Sir Winston Churchill* and the second to the less known (but not less profound) *Mordechai Shapiro*:

I had a feeling once about Mathematics - that I saw it all ... I saw a quantity passing through infinity and changing its sign from plus to minus. I saw exactly why it happened and why the tergiversation was inevitable, but it was after dinner and I let it go.

Sir Winston Spencer Churchill, 1874–1965

Bon vivre, intriguing humor, deep intelligence ... yield a quite good idea of what we might say. But even better what is expressed below.

Symbols are like actors playing on the stage of Math. Like words they are trying to convey meanings and, whatever crazy their combination may appear, it is always possible to gain a hidden message as for hermetic poems. This is why open-mind mathematicians free of prejudices conceived irrationals, transcendental, imaginary ... numbers. The struggle of some mathematicians looks as that for human rights, inside this field there is no privilege and even two opposing truths have the ability of coexistence. Men like Heaviside recognized the power of symbols to decipher the depth of Physics and was always at odds with academic community. What's more...the struggle will be always the same: who is trying to convey novelties and who is worried by changes. Both view coexist and one is unconceivable without the other.

Mordechai Shapiro

Index